Rethinking Science
A philosophical introduction to the unity of science

JAN FAYE
University of Copenhagen

ASHGATE

© Jan Faye 2002

All rights reserved. No part of this publication may be reproduced, stored in a retrieval system or transmitted in any form or by any means, electronic, mechanical, photocopying, recording or otherwise without the prior permission of the publisher.

The Author has asserted his moral right under the Copyright, Designs and Patents Act 1988, to be identified as the author of this work.

Published by
Ashgate Publishing Limited
Gower House
Croft Road
Aldershot
Hampshire GU11 3HR
England

Ashgate Publishing Company
131 Main Street
Burlington, VT 05401-5600, USA

Ashgate website: http//www.ashgate.com

British Library Cataloguing in Publication Data
Jan Faye
  Rethinking science : a philosophical introduction to the unity of science. - (Ashgate new critical thinking in philosophy)
  1.Science - Philosophy
  I.Title
  501

Library of Congress Cataloging in Publication Data
Faye, Jan.
  [Athenes kammer. English]
  Rethinking science : a philosophical introduction to the unity of science / Jan Faye ; [translated by Susan Dew].
      p. cm. -- (Ashgate new critical thinking in philosophy)
  Includes index.
  ISBN 0-7546-0660-0
  1. Science--Philosophy. I. Title. II. Series.

Q175 .F28413 2002
501--dc21

2002018698

ISBN 0 7546 0660 0

Printed and bound by Athenaeum Press, Ltd., Gateshead, Tyne & Wear.

RETHINKING SCIENCE

Science and humanity are usually seen as very different: the sciences of nature aim at explanations whereas the sciences of man seek meaning and understanding. This book shows how these contrasting descriptions fail to fit into a modern philosophical account of the sciences and the arts. Presenting some of the major ideas within the philosophy of science on facts, explanation, interpretation, methods, laws, and theories, Jan Faye compares various approaches, including his own. Arguing that the sciences of nature and the sciences of man share a common practice of acquiring knowledge, this book offers a unique introduction to key aspects in the philosophy of science.

ASHGATE NEW CRITICAL THINKING IN PHILOSOPHY

The Ashgate New Critical Thinking in Philosophy series aims to bring high quality research monograph publishing back into focus for authors, the international library market, and student, academic and research readers. Headed by an international editorial advisory board of acclaimed scholars from across the philosophical spectrum, this new monograph series presents cutting-edge research from established as well as exciting new authors in the field; spans the breadth of philosophy and related disciplinary and interdisciplinary perspectives; and takes contemporary philosophical research into new directions and debate.

Series Editorial Board:

Professor David Cooper, University of Durham, UK
Professor Peter Lipton, University of Cambridge, UK
Professor Sean Sayers, University of Kent at Canterbury, UK
Dr Simon Critchley, University of Essex, UK
Dr Simon Glendinning, University of Reading, UK
Professor Paul Helm, King's College London, UK
Dr David Lamb, University of Birmingham, UK
Professor John Post, Vanderbilt University, Nashville, USA
Professor Alan Goldman, University of Miami, Florida, USA
Professor Joseph Friggieri, University of Malta, Malta
Professor Graham Priest, University of Queensland, Brisbane, Australia
Professor Moira Gatens, University of Sydney, Australia
Professor Alan Musgrave, University of Otago, New Zealand

Contents

Preface		*vi*
1	The Unity of the Sciences	1
2	Reductionism, Emergentism, and Holism	11
3	Explanation	24
4	Interpretation	47
5	Facts	64
6	Methods	83
7	Laws and Rules	114
8	Theories and Models	142
9	Realism and Antirealism	168
10	Beyond the Sciences	201
Index		*215*

To Lisa

Preface

Philosophy of science was founded by the logical positivists as a separate discipline within philosophy. They had great knowledge of science, wrote extensively on mathematics and physics, and wanted to provide a solid ground for doing science – much like Kant did before them. As a consequence the positivists opted for the unity of all the sciences. But their focus was more on the cognitive meaning of scientific theories than on the distinctive scientific practices, and their meta-language was logic rather than ordinary language or mathematics.

The positivists were undoubtedly some of the great thinkers of modernism. But time passed – and a strong reaction against anything that carried a hint of positivism emerged, including the idea of the unity of the sciences. First came the Wittgensteinian turn within the philosophy of science, then the historical turn, and finally the sociological turn. These different movements were forerunners in the ideological confrontation with modernism that culminated in so-called social constructivism and post-modernism. Among them we see an increasing mistrust in a notion of common truth, rationality, representation, method, meaning and origin. All these notions were now to be understood in relation to a language game, a context, a paradigm, a power structure, a society, or a tradition. Feyerabend bit the bullet when he claimed that anything goes within science believing that we could find no general norms of doing science. As much as anybody, he was a philosopher of the post-modern era.

I think we can learn something from the post-modern criticism of modernism, its rejection of absolute and transcendental norms and values, cognitive or otherwise. Nonetheless I also believe that time has come to rethink science and to reintroduce the notion of the unity of science on new ground. The solutions and suggestions that post-modernism and social constructivism have offered are much too extravagant. Even if epistemic values and norms can only be justified with respect to the process of knowledge itself, this fact does not entail that these norms are merely relative to any particular subject, language, power structure, or culture. Instead I call for a revision of the rational basis of science in accordance with the insights we gained from the criticism of positivism and modernism, but still maintaining some trust in the rationality and unity of the scientific enterprise. Thus, the position being advocated here can be seen as a kind of neo-modernism.

For the most part this book is a translation of a work I recently published in Danish under the title *Athenes kammer. En filosofisk indføring i videnskabernes enhed*. Only minor additions and corrections have been made except in Chapter 3 which has been considerably enlarged. I wish to express my gratitude to the Danish publisher, Morten Hesseldahl, Høst & Søn for allowing me this English edition. My thanks also go to Susan Dew for having struggled successfully with the translation, and no less to my friend and colleague, Henry Folse, who with characteristic generosity has commented on the entire manuscript and thereby helped to improve it greatly. I am also indebted to the Danish Institute for the Advanced Studies in Humanities for funding making this translation possible.

Copenhagen, June 2002

1 The Unity of the Sciences

Science is one of the cornerstones of modern society. Indeed, it is true to say that the development of science since the Renaissance has been crucial to the emergence of modern society. Today vast sums of money are channelled into science in the expectation that it will yield solutions to specific problems, and by so doing enhance the conditions of human life and further technological progress. Science enables us to combat disease, furnishes us with knowledge about the structures of chemical substances, thereby enabling us to manufacture new materials; it studies the forces of nature in order to harness them for the production of energy and heat. But quite apart from these practical aims, science satisfies the common human aspiration to acquire systematic knowledge of nature and society and to arrive at an understanding of our own peculiar nature, culture and history.

As determined by its object, science is partitioned into natural science, social science and human science. Briefly, the natural sciences seek to attain knowledge about both inanimate and animate nature. They encompass such subjects as astronomy, physics, chemistry, geology, meteorology, biology, botany and zoology. The social sciences, on the other hand, study society or sectors of society and their subdivisions; its compass takes in states, nations, political parties, organizations, institutions and the family. Among the various disciplines in this area we find economics, political science, sociology and anthropology. And lastly we have the human sciences. These comprise a very large group of varied disciplines whose focus is the human person *qua* thinking, willing and acting subject, and human cultural products ranging from linguistic utterances, texts and works of art to myths, utensils, tools and clothing. They encompass human history too in all its dimensions and aspects. The humanities include such subjects as psychology, linguistics, art history, film, comparative literature, ethnology, archaeology and the history of ideas.

It is important in a democratic society that citizens are informed about what science is in a general sense, what distinguishes the particular sciences and marks them off from one another. Even if, in the final analysis, it is the researchers themselves who define their subject, it is important that the wider public be capable of entering into critical dialogue with them.

Even more important is that scholars and scientists not only know about the subject content of their own discipline but also have a fairly clear idea of what science is – what marks out science as a species of knowledge. For even if a researcher knows a good deal about physics, biology, economy or philology, he or she need not necessarily know very much about what science is in a generic sense, or how it is practised. Habitual engagement with the problems of a particular subject takes the form of practical knowledge comparable to the ability to ride a bike. Most people can ride a bike without being able to explain how they do it. And just as it is not crucial for a cyclist to be able to offer an account of the physical laws involved in keeping one's balance, it might seem of as little importance that the researcher know what science is in order to do science. Nevertheless, knowledge of the nature of science is desirable for researchers as well as laypersons if they are to be capable of forming a general conception of the value of scientific findings, or if they are to know the difference between good and bad research, or be able to establish an evaluative basis for the allocation of government funding to research etc.

Notwithstanding the fact that the sciences and the humanities are severally concerned with very diverse subject areas, and the objects of study within the individual areas are very varied, it is noteworthy that we still refer to as 'research', both individually and collectively, the activities attaching to the systematic investigation of apparently heterogeneous phenomena. Physics, economics, history, and literary studies are all sciences or fields of research. But are they so classified at the dictate of tradition, or are there common features in the practice of research that link these disparate subjects and make them 'sciences'?

Physics and literary studies seem to be as far apart as can possibly be imagined. Many scholars and natural scientists have thus questioned the notion of a shared basis for the practice of these two disciplines. It is often said of the natural sciences that they are concerned with objective facts and that they seek to uncover the universal laws that explain the individual phenomena we observe. It is otherwise with the human sciences. Here it is very much a matter of subjective interpretation whose correctness cannot be objectively established and which is, accordingly, to some extent dependent on the researcher's own norms and values.

Listed below are some of what are often claimed to be immediately identifiable differences between the humanities on one hand and the natural, and in part the social, sciences, on the other:

1. The non-human versus the human.
2. Physical matter versus the thinking subject.
3. Collective versus individual.
4. Explanatory versus interpretive.
5. Nomothetic versus idiographic, *i.e.* concerned with universal laws versus individual events.
6. Value-freedom versus value-commitment.
7. Objectivity versus subjectivity.

Human beings and their thought, attitudes, values and views of life are the very things with which natural science does not occupy itself. In a word, it is not concerned with the understanding of subjectivity. That is precisely the preserve of the humanities. In consequence, the individual and all that is peculiar to individuals are pivotal to the human sciences. Here, it is said, the aim is to understand human subjectivity – its signs, symbols and agency – as items that cannot be explained or described in an objective manner.

Nothing on the list suggests any significant similarities between the various fields of research. This makes it difficult to understand why the various domains are considered as parts of a single enterprise. However, if this observation is well founded, it raises an interesting problem. For if there is no common denominator among the sciences and other fields of research viewed collectively, what makes it possible to reject the idea that astrology and parapsychology constitute science? This issue is known as the problem of demarcation. Is their exclusion simply grounded in tradition? It is one of the tasks of the philosophy of science to address such questions.

Now and again one hears that, for all their apparent differences, the sciences and the humanities do indeed have sufficiently many features in common to enable them to be viewed as a collective endeavour aiming at the attainment of knowledge and insight into everything knowable. The vision of the unity of all the sciences concerning nature and man is not new. Indeed, it can be traced back to antiquity. We meet it in the debate about *physis* (nature) and *nomos* (law or custom) – in the question as to whether these are contraries, or whether the one is reducible to the other. It springs basically from the fact that, in his or her search for structure and uniformity underlying the manifold of phenomena, the human subject is guided by the idea that the contents of the world constitute a coherent whole with a common origin in one cosmos. Our conceptions of the world must thus seek to reflect this coherent whole if they are to lead to truth.

Aristotle's animistic conception of nature was the expression of just such a cosmic idea in that it sought to understand one part of reality by viewing it on analogy with another part. We see a resurgence of this vision in the Renaissance studies of nature and human beings in their concrete and universal contexts. In the twentieth century in particular, the philosophical orientation called logical positivism sought to show that the sciences are not essentially different from each other. The way the positivists developed this idea, however, was far from unproblematic. To establish the unity of science they ignored those important features of the human and social sciences that set them off from the natural sciences. Had their prescriptions for establishing the unity of science been followed, the bulk of research conducted within the humanities would not have qualified for the appellation 'science'. Traditional critiques of positivism thus turned on the extent to which they blankly applied the scientific ideal of *natural* science to those sciences which ontologically, methodologically and epistemologically, were fundamentally different.

The success of this critical reaction must be said to be qualified. It led to the widespread conception at universities that there exists a vast gulf between the natural sciences on the one hand and the human sciences on the other. This perceived divide comes to expression in the social sciences in the form of a discussion about the extent to which these sciences of society can be, and should be, nomothetic or idiographic: whether the social sciences should describe the regularities underlying the manifold of social phenomena or whether they should instead seek to get to grips with what is peculiar to each individual state, nation, party, system, etc. by offering an interpretation of the conventional rules and meanings that are characteristic of such institutions. The general tendency is to say that natural science and also, in part, social sciences are nomothetic and explanatory while the human sciences are idiographic and interpretive. This disparity is owed, it is claimed, to the important fact that nature comprises causal processes and natural kinds while the human person is an intentional being with a cultural output underpinned by conventions, norms and values. The standpoint has subsequently been used in support of the claim that research is driven by a variety of different aims, according as it targets nature or human beings. It is in fact a common assumption that natural science and the human sciences stand in direct opposition to each other. One result has been that natural science has come to figure as an object of study for analytic philosophy while the knowledge acquired in the human sciences has become the province of continental philosophy, not least of philosophical hermeneutics.

There seem, however, to be important differences within the humanities which place a query against the plausibility of their being susceptible to just one method. The human sciences may be grouped into:

1. Those that study the human subject *qua* thinking, feeling and willing agent and the mechanisms underlying such psychological dispositions.
2. Those that study the products of human agency such as linguistic utterances, art, literature, music, tools, clothing and utensils.
3. Those that study the chronological order of a set of data or events.

Is it utterly obvious that hermeneutics is the right method for each of these areas? Does the study of human language acquisition lend itself to the same methods, as does the development of portraiture in the history of painting?

To this challenge Martin Heidegger and Hans-Georg Gadamer reply on behalf of hermeneutics by saying that it is not a discipline exclusively concerned with the understanding and interpretation of written documents and speeches. The object of understanding is not merely an individual's communicative output or psychology. The meaning of a text always extends beyond the intention of the author or the partner in conversation. The focus of hermeneutics is more fundamental than any dichotomy of natural science (explanation) and the human sciences (understanding). Hermeneutics is the study of the understanding of nature, culture and the life of the mind. Consequently, hermeneutics harbours within it the idea of the unity of science.

Gadamer's position may be summarized briefly as follows: Truth is something other than method. Hermeneutics is a universal methodology. Understanding, explanation and application cannot be prised apart. Every interpretation involves application: Gadamer here aligns interpretation with the interpretation of laws and rules. Every text, every work and every theory imposes constraints on its own interpretation once the interpreter understands that it involves herself. We cannot, for instance, separate the horizon of the author from that of the interpreter. There must necessarily be some overlap between them for an understanding of the text to be possible. Prejudices, which are really 'pre-judgements', comprise all the beliefs and attitudes constituent in the horizons of understanding that we bring to the process of understanding. We acquire a fresh understanding whenever our pre-judgements fail to fit with what we are trying to understand. The work's effective history becomes the sum of interpretations and accommodations made over time. Understanding comes about through real conversation. Gadamer speaks of a fusion of horizons occurring when the understanding that finds expression in the work and that of the interpreter fuse. Finally, he claims that a grasp of completeness is a condition of understanding. This is especially manifest in the understanding of classic or superior texts to whose authority we should yield. Authority, he avers, is not the antithesis of freedom and reason.

Precisely this last point is the target of Habermas' vigorous criticism. His critical theory constitutes an attempt to understand the unity of science in terms of communicative action. He points out that all translation may be construed as a hermeneutical process in which we 'translate' when we fail to understand what we seek to understand. Language is dialectical: it comprehends its own preconditions. The rules of language do not simply supply us with usage but also with the rules for their own understanding and interpretation. Habermas concurs with Gadamer in what concerns the latter's opposition to objectivism. There is no such thing as the determinate objective reconstruction of a text or a historical event. He also agrees with Gadamer that understanding and use cannot be separated. We reach a correct interpretation through agreement, and agreement can arise in two ways: (i) we can bring our beliefs and attitudes (the horizon of understanding) into line with those of the text (textual authority), or (ii) we can bring the text's contents and attitudes into line with ours (relativism). But neither option is satisfactory. A third option is that of distinguishing between understanding and misunderstanding (between a good and a bad interpretation). There are beliefs that enter into our horizon of understanding but which distort our understanding and which, consequentially, can only be understood through *reflection* in which a critique of ideology is implicit. Habermas introduces the notion of unconstrained communication, free of illegitimate authority.

Although Gadamer and Habermas aim at formulating a general philosophical theory of the unity of scientific knowledge they fail fully to succeed, because the thrust of their analysis is coloured by a focus on the understanding of texts and communicative action. This overlooks the fact that nature is silent. It does not speak to us, but we apply ourselves to addressing it. We cannot achieve a fusion of horizons with nature. If nature appears to answer, it is merely the echo of our own questions we hear, reverberating down to us from their Olympian heights.

This book seeks to reopen the philosophical debate between the natural, the social and the human sciences. It attempts this via an analysis of current practices in each of these areas to see whether these many disciplines do in fact have features in common. Scientific practice comprises all the activities that underpin the current body of knowledge within a particular science or field of study, and which are governed by a set of epistemic and methodological rules. I shall be arguing that the idea of the unity of science does indeed manifest itself in the work done in the natural sciences, the social sciences and the human sciences. However, this fact is a fact not readily apparent to the active researcher, since the scholar, say, and the social scientist, have only a slight acquaintance with natural science, and the natural scientist knows very little about the social and human sciences. This is understandable, however, considering that the point of departure, namely, the subject matter, varies enormously from one subject to another.

A glance at the history of the humanities is enough for the affinities between the natural and the human sciences to be revealed. The Renaissance marked a resurgence of interest in the study of the human subject. It was no longer adequate to characterize human beings by reference to their status as God's creatures: they are in possession of emotions and intellects; they have language and the freedom to create and understand. The Bible was translated into the vernacular and many philosophical and scientific texts were translated from Greek and Arabic into Latin. Such translations required textual understanding and an assessment of the chronology and authenticity of the extant manuscripts. Later Newton's classical mechanics ushered in a new era in the human sciences. In the Enlightenment the human sciences developed the distinction between empirical methods and theory – empirical data formed the touchstone of theory, a theory had to correspond to experience. If we look at, say, the progress of linguistics we see that the initial step consisted of the identification of parts of speech and the classification of words in other respects: just as in biology species are identified and their evolution traced, so too the aim in linguistics became that of tracing the evolution of individual words. This phase is known as the historico-classical tradition. The next phase was structuralism, which rejects a diachronic approach in favour of a synchronic approach. Language is now understood as a semiotic system – and the following phase produces generative grammars where a distinction is drawn between depth and surface grammar. Literary studies have traced a similar process of development and a range of highly diverse theories has appeared. Initially biographical theories held sway, until succeeded by the generic-comparative, structuralist, psychoanalytic, Marxian, and most recently by deconstructivist theories. The scholar to no lesser degree than the scientist has developed theory to explain and understand the world. The diversity of theories in the human sciences is comparable to that in the natural sciences. Here, too, a host of theories have been elaborated which researchers use in seeking to explain their findings.

Every scientific theory is predicated on a considerable degree of abstraction from and idealization of concrete reality. By defining their field of enquiry the individual sciences designate their own materials, for they focus on certain features while quite ignoring others. The distinctly generic analysis of the sciences offered by the philosophy of science rests likewise on an abstraction and idealization of the individual sciences. Many levels of description and understanding may be fitted into the space separating the primary object from philosophical reflection, each of which creates their own intellectual genre and mediates their own practice. If we turn our gaze once again to fiction we will observe a series of nested epistemic strata:

- Philosophy of science and humanities
- Science studies
- Text studies
- Literary studies
- Criticism
- Poetics
- Fiction

Each of these levels is made the object of description and analysis at the next level up. At each successive level, wholly new concepts are introduced to mediate the understanding and explanation of lower levels – without the introduction of such novelty there would be no new level.

Fiction comprises concrete individual works. These include novels, plays, short stories, anthologies of poetry etc. Poetics may be understood as the body of rules having application to the art of poetry – guidelines issued by philosophers or authors for the writing of literature. Criticism then treats of literature and poetics in the form of reviews and essays that are designed to introduce potential readers to the work. The critic addresses the work against the background of his readings of other individual works, and focuses normatively and argumentatively on the work's literary and aesthetic qualities, but without basing his assessment on a theoretical or systematic treatment of the work.

By contrast, literary studies constitute a theoretical and systematic discipline. By using abstract concepts, general assumptions and methodical enquiry this discipline seeks to grasp literature, poetics and criticism *qua* literary and aesthetic products. Text studies address texts of all kinds. Indeed, poetics, criticism and literary studies are themselves producers of texts that are the material of textual theory. For the purposes of text studies, literature is a text on a par with pulp fiction, advertisements and political, religious and scientific texts. Science studies seek, in a very general way, to articulate a precise determination of the various scientific subject areas and their interrelations, in order to produce a theoretical analysis of the development of science that takes off from the history of science, the sociology of science and the psychology of knowledge.

Lastly, we come to the philosophy of science and the humanities, which reflect on the epistemic and methodological conditions governing these subject areas. The philosophy of science and the humanities introduce a suite of philosophical concepts through which sciences of nature and the sciences of man may be described and understood. These are concepts such as 'truth', 'empirical adequacy', 'observation', 'fact', 'method', 'deduction', 'induction', 'reliability', 'verification', 'falsification', 'explanation', 'interpretation' and many more that we shall encounter as we proceed. The philosophy of the arts and the sciences is, in line with other branches of philosophy, not tied to any particular method – in contrast to the sciences, philosophy is not identified with particular methods. Philosophy is not a science. The humanities seek to explain and interpret a specific range of phenomena making a systematic collection of empirical information their point of departure, but philosophy's stance with respect to its subject matter is critical and analytical and rests on no such systematic collection of empirical data. Instead, philosophy frames its own methodological problems. It could hardly be otherwise. The philosophy of science seeks to identify the conditions governing *all* scientific knowledge, including the use of the methods applied by both the empirical and interpretive sciences. Evidently, then, if its results are to have validity, philosophy cannot use the same methods as those whose conditions of legitimacy it investigates.

The picture that present-day philosophers of science paint of the theory and practice of natural science is different from and more realistic than that once presented by positivists, although the latter is still regarded by many scholars and natural scientists as displaying, to some extent, the face of natural science. The new picture is at some points closer to the portrait hermeneuticists have drawn of the humanities. But that portrait, too, fails fully to capture the likeness of its subject. This goal will only be achieved by our presenting a version of humanities less aloof from present-day conceptions of the natural sciences. Both these fresh portraits prove on comparison, and not surprisingly, to contain the features that the picture of the unity of science brings together.

References

Ayer, A.J. ed. (1959), *Logical Positivism*, Glencoe, Ill.
Gadamer, H.-G. (1960), *Wahrheit und Methode. Grundzüge einer philosophischen Hermeneutik*, J.C.R. Mohr, Tübingen.
Habermas, J. (1968), *Erkenntnis und Interesse*, Suhrkamp, Frankfurt am Main.
Von Mises, R.E. (1939/1951), *Positivism, A Study in Human Understanding*, Harvard University Press, Cambridge Mass.

2 Reductionism, Emergentism, and Holism

The notion of the unity of science is ambiguous. The expression 'the unity of science' can mean (a) that there are common features attaching to all the objects of which science treats, or (b) that there are features common to the ways through which we attain knowledge of these objects. We must distinguish between these two senses. The present chapter focuses on the first sense, and I shall seek to show that the unity of science in this sense is untenable. The second sense is the topic of the remainder of the book.

A number of scientists maintain that physics is the most fundamental of the sciences, capable of explaining all the others. And among those who have not voiced this claim many tacitly subscribe to it. The idea is that we can understand how bigger things work if we understand how smaller things work. In physics the laws governing bigger things rest on the laws governing smaller things. Chemistry builds on physics, biology on chemistry, psychology on biology, and, finally, sociology on psychology. The fact that it has not yet proved possible to explain the connections in detail owes simply to present human ignorance and the vast complexity of big things.

Sound scientific knowledge leads, then, to a unity informing our conception of the world, whether it concerns nature, society or human beings. This claim has found support in a variety of ontological and epistemological arguments. However, in order to evaluate them we must know what is correct scientific knowledge and what the criterion for unified knowledge. It must be possible to state the features that show us where this conception of unity comes from. It must also be possible to show that such a conception of unity is always conducive to sound scientific knowledge.

The logical argument

It would seem to be possible to formulate the requirements for unifying scientific knowledge in the following terms:

1. First principles
2. Unification
3. Reduction
   (a) nomic reduction (simplicity/generality);
   (b) conceptual reduction (simplicity/generality).
4. 'Ockham's razor'

If the notion of the unity of science is to be tenable it must be possible to trace the branches of enquiry back to several first principles from which all other scientific claims and assumptions may be derived. The principles would be the simplest and the most general: simple in the sense that they are not composite and general in the sense that they would encompass and exhaust that subsumed under them. Such first principles unify all systematic knowledge through all other principles being derivable from them. This is also known as the procedure of 'reducing' more specific principles to the most basic ones.

This reduction may be carried out in two ways: Scientific unity could be achieved through the *reduction of laws* if it could be shown that secondary laws can be derived from more general laws because the latter have more generality than the former. Unity could also be achieved through the *reduction of concepts*. A set of specific and highly specialized concepts reduces to simpler and more general concepts. This might be accomplished by the introduction of a bridge law that establishes the identity of two distinct entities.

Conceptual reduction implies something other and more than subordination to a concept of greater generality. To illustrate: the concepts 'table', 'chair', and 'bed' might be ordered under the concept 'item of furniture' which in turn, together with 'cutlery', 'kitchen utensils', 'carpet', 'painting', and 'bric-à-brac' may be categorized under the concept 'household effects'. The example also makes clear why 'table', 'chair', and 'bed' cannot be severally reduced to the concept 'item of furniture', etc. For the concept 'bed' to be reduced to the concept 'item of furniture' the expression 'item of furniture' would have to denote nothing beyond what the word 'bed' denotes. But 'item of furniture' covers many things apart from 'bed'. A concept can only be substituted for another concept if both have the same extension. It is not necessary that they express the same meaning.

Ernest Nagel is probably the philosopher of science who has most systematically sought to show how nomic and conceptual reduction could be possible in science. In his major work *The Structure of Science* from 1961 he sets out two conditions that must hold in cases where a theory T_2 is to be reduced to another theory T_1:

1. *The condition of connectability*: For every theoretical term 'M', which occurs in T_2 but not in T_1, there is a theoretical term 'N' which is constructable in T_1 but not in T_2 such that:

(B) for all objects x, x has M, if (and possibly only if) x has N,

where (B) is called a '*bridge law.*'

2. *The condition of derivability*: It must be logically possible to derive all T_2's theoretical laws from T_1's theoretical laws in conjunction with bridge laws.

Consequently, we have:

(X) $(L_1 \& B) \supset L_2$ where L_1 is the set of laws in T_1 and L_2 is the set of laws in T_2,

where $\supset$ is the logical implication sign which reads 'if.., then..' or 'entails'. If these conditions are satisfied, argues Nagel, then T_2 has been reduced to T_1.

By 'bridge law' is meant a principle that lays it down that a given phenomenon is identical with some other specified phenomenon. When we say that 'water' is identical with the chemical compound 'H_2O' we make reference to a bridge law. It is a condition of such a law that 'water' and 'H_2O' have the same extension, *i.e.* the words refer to the same substance and only to that substance. But the words do not mean the same; they do not have the same *intension or semantic content*. By the word 'water' we understand something other than H_2O. But we still say that both terms designate the same substance.

Nagel's two requirements for reduction appear to be very precise. But only rarely can they be satisfied. In the first place, the condition of derivability scarcely holds between the theories Nagel imagined capable of being reduced, the one to the other. Logically speaking, the condition of derivability is valid if L_2 is not false when L_1 and B are true. If we imagine that L_2 are Newton's laws, and they are false, it becomes difficult to understand how they might be logically derivable from true laws – say, those of quantum mechanics.

But even if it is maintained that Newton's laws are not false, it is still impossible to derive them from the theory of relativity or quantum mechanics. That impossibility is due to the fact that scientific theories determine the meaning of the theoretical terms in a way that fixes their meaning in relation to other terms figuring in the theory in question. So even if the term 'mass' has the same extension in both theories, it does not have the same intension, and so the one theory is not derivable from the other. This creates problems in relation to the condition of connectivity.

John Kemeny and Paul Oppenheim have advanced criticism along these lines. In some cases it is possible to carry out a reduction by bridging two theories such as, for example, thermodynamics and statistical dynamics. But as these two philosophers rightly point out: even if we have a well-functioning reduction between those two theories, we are still only able to identify a few obvious examples of bridge laws because the meanings of the terms involved cannot be fixed independently of the theory of which they are a part.

In light of this criticism, Kemeny and Oppenheim propose instead that a reduction of T_2 to T_1 goes through if:

1. T_2 contains terms not found in T_1.
2. All empirical data explained by T_2 is also explainable by T_1.
3. T_1 is at least as well systematized as T_2.

But here too we run into serious difficulties. For according to (1) T_1 is semantically distinct from T_2 and (2) above is to ensure that T_2 can be reduced to T_1. However, it is possible for us to have two incompatible theories that are empirically underdetermined. A theory T is *empirically underdetermined* if and only if there is an alternative theory from which precisely the same observational consequences may be derived as are derivable from T. Incompatible theories can, then, be empirically indistinguishable because they explain the same observational data while at the same time being theories of which it must be universally conceded that they cannot be reduced, the one to the other. In principle an infinite number of hypotheses may be formulated on the basis of a finite number of data. Plainly, this means that (2) is not, after all, able to ensure that T_2 is reducible to T_1.

Moreover, other objections can be raised against (3). There are a number of examples in the history of science showing that the explanatory theory T_1 does not need to be at least as systematic as the theory explained, T_2, irrespective of whether the systematization has to do with the number of laws and principles or with a well-ordered and well-structured composition. It must thus be concluded that Kemeny's and Oppenheim's conditions for reduction also emerge as less than unproblematic.

The idea behind nomic and conceptual reductions rests on Ockham's razor, so-called – the principle that states that no more laws or concepts should be used than are necessary since any such will be ontologically superfluous. By working with the most simple and general concepts one can rest assured that no more entities and properties than are needed are being postulated as existent in the world. The difficulties attaching to setting up criteria for reduction would appear to suggest, however, that our multi-faceted theories of the world do indeed correspond to its variety.

As already indicated, scientific knowledge must satisfy certain requirements in order to be perceived as offering support to the idea of unity of science. The proponents of the unity of science imagine that it is possible to succeed in identifying first principles and that, with that foundation in place, we would be able to reduce the social sciences to psychology, psychology to biology, biology to chemistry, chemistry to physics and ultimately to derive physics from some hoped for theory of everything. Of course no one maintains that science *today* can reduce the sum of scientific knowledge to such first principles or that there actually exist fully elaborated reductions of the different sciences – with the derivation of chemistry from physics as the possible exception. But its proponents will reply that science is still in its infancy, and that just because such reductions have not *yet* been carried through this does not mean that they will not be achieved *at some future time*. There is, then, good reason to enquire into whether such conditions can be formulated *at all* and – if they can – whether they would be conductive to sound science.

The cosmological argument

Everything in the universe took its beginnings from the Big Bang. Physicists claim to be able to explain how the universe evolved and how it is that matter is homogeneously distributed in space. The laws of nature as we know them today appear to have been in operation already at the inception of the universe, or, if it is claimed that they assumed forms different from those met with today owing to symmetry breaking, we are still able to offer some explanation of how these forms evolved from earlier ones. But it means, then, that present conditions are continuous with earlier conditions. This makes it difficult to understand how there come to be things in the world which can only be described using quite other laws and concepts than those of physics, simply because physics is the science that accounts for the laws and conditions obtaining when the universe began.

Paul Oppenheim and Hilary Putnam once argued for the unity of science on the basis of just such a cosmological conception. They set up two premises for their argument. One they called the 'evolution hypothesis' and the other the 'hypothesis of ontogenesis.'

The principle of evolution: For every level of organization, there is at any given point in evolution this level, but no higher level of organization.

The principle of ontogenesis: For any system at a given level of organization, there is a point in its development at which the system did not exist, but all its elements existed.

According to the first principle there was at a given time things of level n but no things of level $n+1$. At a point in time, that is, when there were elementary particles, there were no atoms, and at a point when there were atoms there were no molecules. The second principle says that for any particular object at level n, there was a time when it did not exist, but where its parts existed at level $n-1$ – parts from which it developed or by which it was caused. On the basis of these two principles Putnam and Oppenheim concluded that everything at a given, specified level is ontologically dependent on the level immediately below it. Each and every thing and its properties can be explained on the basis of its parts and the properties of these parts and their relations. Such an explanation is called an *ontological reduction*.

There are two kinds of ontological reduction: *eliminative* if x is reduced to y with x being eliminated by y, and *non-eliminative* if x is reduced to y without x being eliminated by y. Eliminative reductionism must of course explain why x *appears* to be an entity even though, ontologically, it *is* not, while non-eliminative reductionism needs to explain how x can be reduced to y without ceasing to exist. Putnam and Oppenheim's argument may be regarded as a defence of non-eliminative reductionism.

Ontological reductionism is vulnerable to several lines of criticism. The most radical attack would be to show that at each level there are laws and natural kinds that are not susceptible of explanation in terms of the level below it. At the same time, there is room for disagreement with regard to the extent to which these laws and classes may be explained by reference to the level above. Those opposing ontological reductionism are called 'holists' and 'emergentists'. The difference between *emergentism* and *holism* is precisely that between saying that ontological independence obtains between the individual levels and saying that the lower level is ontologically dependent on that above it. The holistic standpoint holds that features of the whole cannot be explained by reference to the parts but that features of the parts can be explained by reference to the whole. Emergentism, conversely, denies that either the whole or the parts explain the several levels.

One way of repudiating ontological reductionism is to follow Jerry Fodor in arguing that the divide between psychology and neurophysiology cannot be crossed. A discontinuity obtains between these organizational levels: the levels are emergent in relation to each other. Psychological and neurophysiological phenomena belong to fundamentally *distinct ontological types*. Mental states are *not* reducible to physical states through the use of bridge laws.

Fodor argues that the same mental state may be realized on the basis of many different physical states. Take the thought of the Garden of Eden, for instance. That thought will be the same as held in the minds of both Adam and Eve, but it may be related to different neurophysiological processes in their respective brains. No one-to-one relation can be established between phenomena of a mental type and a physical type. In consequence, it is not possible to establish a putative law connecting the mental and the physical. The following is a brief summary of the salient points in Fodor's argument.

1. There exists a special level of organization in nature in which there is a discontinuity between phenomena of different kinds.
2. Therefore, there is a theoretical discontinuity between the different sciences.
3. Natural kinds on individual levels prevent conceptual and nomological reduction to a lower level.
4. There exist no bridge laws linking the mental and the physical.

The natural kinds referred to by Fodor are the entities and the matter that plainly occur in nature and whose existence is not contingent on practical human activity or theoretical knowledge. These include gold and iron, lions and zebras, atoms and molecules. The world is already, from the hand of nature, divided into 'kinds' or 'classes' and it is up to us to ascertain their objective existence. We shall later return to the questions relating to natural kinds but let us for the time being assume that we can give credence to them without conceptual problems.

Natural laws are what determine natural kinds since they govern the individual members that make up the class or rule between various items that belong to different classes. If we can establish a causal connection between a unique type of mental phenomenon and a unique type of physical phenomenon we shall have succeeded in establishing a law governing these phenomena. But this is what we cannot do, Fodor points out, because the connections between them are not uniform.

We are, by the same token, unable to employ bridge laws to achieve a conceptual reduction. To create a bridge law between the mental and the physical it must be true that 'The thought of the Garden of Eden' and a particular neurophysiological process 'NP' refer to one and the same thing. It requires that:

For all organisms o, o has M if (and possibly only if) o has (NP_1 or NP_2 or NP_3 or ...$NP_{28.374}$); and all NPs belong to the same type of NP,

where 'M' is a mental state. But the NPs would not all be identical inasmuch as they do not share any one determinate feature. The objection builds on a functional characterization of the mental state M.

Responding to Fodor's challenge, Jaegwon Kim asserts that a supervenience relation may still obtain between the mental and the physical. On the one hand he agrees with Fodor that the mental cannot be reduced to the physical but on the other he wishes to avoid postulating entities that do not exist. We can easily imagine a world constituted exclusively of physical entities but in which these display different types of properties. So he introduces the notion of 'supervenience' according to which one phenomenon is said to 'supervene' on another if there is a relation of asymmetrical dependence between them. But it has to be a relation of a highly distinctive kind. It is not enough to be able to say that there is asymmetrical dependence. Such dependence might be causal, perhaps allowing us to establish a causal law between the two sets of phenomena. The problem Kim seeks to solve can be put quite simply: How do we construct a concept (B) of asymmetrical dependence for which the relation of dependence is:

1. Non-causal (it does not require law-like regularity).
2. Non-reductive (allows of a one-many relation).
3. Ontologically significant.

that enables us to defend non-causal, non-reductive physicalism?

This is where Kim's notion of supervenience enters. There holds, he argues, a relation of asymmetrical covariation such that mental properties (considered as a family M^*) may be said to *supervene* on physical properties (considered as a family N^*) if and only if the following relation holds between these properties:

> Necessarily, for every object x and y, if x and y have exactly the same N^*-properties they also have exactly the same M^*-properties; and if x and y differ from each other with respect to M^*-properties they also differ from each other with respect to N^*-properties.

In other words, what 'supervenience' means is that the mental supervenes on the physical if and only if asymmetrical covariation obtains between mental and physical phenomena: if the mental varies so does the physical. Conversely, the physical can vary without the mental doing so. Further, the two mental phenomena will be the same if the underlying physical phenomena are the same.

But supervenience is necessary only for (B), it is not in itself sufficient for (B). Asymmetrical *covariation* does not yield asymmetrical *determinative* dependence. That asymmetrical covariation obtains between two domains does not mean that either of these domains is necessarily determinative of the other. Recall Leibniz's doctrine of pre-established harmony between monads – they are aware of each other's behaviour without being interconnected. And even if there existed such determinative dependence in our world, it still remains for science to show what it consists of. Is it neurological, causal, semantic or something quite different? It is easy to conceive of its varying from one domain to another. It is up to the individual sciences to establish whether asymmetrical covariation or determinative dependence is to be found and to determine how it can be more precisely characterized.

Plainly, then, ontological reductionism offers only a weak foundation for the notion of the unity of science, if Fodor's or Kim's arguments are accepted. For even if it is supervenience that is defended rather than emergentism, ontological monism has not been vindicated until it has been shown that the determinative connection between the several levels rests on a causal connection. This, however, would appear to be only a remote possibility. So if the unity of science is to be something other and more than merely a heuristic and regulative idea and also express a condition of sound scientific practice, it must seek support elsewhere.

The epistemic conception

There is not only an ontological version of reductionism. Epistemological reductionism was another version that was to constitute a focus for logical positivism. It starts from what we can know. We cannot comprehend the world in other terms than those delivered by our sensory modalities. We cannot know anything about the world as it is in itself. In consequence, we must reduce our theoretical descriptions of the world to descriptions of sensory experience. The articulation of the notion of the unity of science starts from empiricism's claim that if science is to have the status of reliable and certain knowledge, it must take its beginnings from what is given in immediate sensory experience, which is to say sensory input or sense-data. Such sensory experiences are not simply the source of all knowledge – they are also indubitable. Our beliefs about them are incorrigible. Only if all hypotheses reduce to statements about perception do they qualify for the status of certain, scientific knowledge. This, then, in capsule form, is the positivists' proposal as to how the warrant of the truth claims of science may be demonstrated.

The positivists were later to depart from neutral sense data as the foundation of knowledge. Instead they came to ground their ideas on the physicalist notion that all scientific statements should be reducible to an entity language capable of satisfying a publicly agreed constraint and thereby come to refer to observable entities. The positivists drew a distinction between the language of observation and the language of theory. The language of observation contained terms for only those phenomena that could be observed whereas the language of theory contained words for entities postulated by theory. Observational terms and propositions acquire their meaning from a correlation between words and visible things – so-called ostensive definitions – while theoretical terms receive their meaning from being translatable into observational terms. At the same time observational statements are, in contrast to theoretical statements, truth-bearers.

The schema below sets out the central points of the positivists' conception of language.

O-language	T-language
• The meaning of O-terms is determined through a correlation with things and their properties.	• The meaning of T-terms is determined indirectly via a translation to O-terms.
• O-terms refer.	• T-terms do not refer.
• The meaning of O-statements is determined by observable truth conditions.	• The meaning of T-statements is determined by their systematic role within the theory.
• O-statements are truth evaluative or bearers of truth.	• T-statements are not truth evaluative or bearers of truth.

Naturally, the positivists' conception of language bore on the broader constraints they imposed on any notion of the unity of science: every epistemically meaningful statement must be directly empirically verifiable or at least reducible to statements that may be immediately validated by experience. Their objective was to have such a verification principle rid science of empty metaphysical speculations, which, according to the positivists, was not the expression of meaningful epistemic claims. To achieve this, the following constraints on the principle were formulated:

(a) That it be formulated using precise concepts that could be captured in an extensional logical language;
(b) That it should capture as epistemically meaningful all the sentences which play a significant role in natural science; and
(c) That it should exclude sentences that are the expression of metaphysical speculation.

By so doing the positivists excluded everything in the social and human sciences that cannot be expressed in a material object language and which, in consequence, does not satisfy the verification constraint. Into this category fall the intentionality of agency, mental states and various forms of meaning and understanding. Further, the principle robs sentences about values and evaluations of epistemic content. Now meaning, understanding and values are the *differentiae* of the objects studied by the social and human sciences. Had the informing idea of positivism been followed it would, in its eagerness to secure for science a firm and indubitable foundation, have stifled all research except that in natural science.

Very few present-day philosophers of science accept as a whole the positivist programme and indeed the idea that it should be possible to realize the idea of the unity of science on the basis of the assumption that epistemic goals such as truth, simplicity, and coherence, find expression in the social and human sciences as they do in natural science must be rejected outright. If the ontological domains are different, the epistemic interests and epistemic aims must likewise be formulated differently accordingly.

The methodological conception

The unity of science understood as the claim that there is a common denominator for the objects of scientific knowledge lacks both convincing empirical and philosophical support, and is in outright conflict with an interest in the study of the human subject and its self-expression in art and culture. On the other hand, the notion of the methodological unity of science affords us purchase on the conduct of science. The latter represents a different understanding of the concept. Here the idea is that certain methodological prescriptions are generic to the sciences and the activity we call 'science' is characterized by its employment of the same methods. So methodological unity of science does not maintain that one ontological domain can be reduced to another, that laws or the realm of entities within a domain may be described in terms of laws or the realm of entities within another domain. The position is, rather, that despite the important differences that obtain between two ontological domains such as physical nature and human culture – and by implication the differences between the explanations through which we seek to ground our knowledge – every scientific investigation of these domains must be conducted on the basis of a common set of methods. It is this methodological conception of unity for which I shall be pleading in the remainder of this book.

References

Fodor, J. (1975), *The Language of Thought*, Harvard University Press, Cambridge Mass.
Kemeny, J. and Oppenheim, P. (1956), 'On Reduction,' in *Philosophical Studies 7*, pp. 6-19.
Kim, J. (1993), *Supervenience and Mind*, Cambridge University Press, Cambridge.
Klee, R. (1997), *Introduction to Philosophy of Science Cutting Nature at its Seams*, Oxford University Press, New York and Oxford, Chap. 5.
Nagel, E. (1961), *The Structure of Science*, Harcourt, New York, Chap. 11.
Oppenheim, P. and Putnam, H. (1958), 'Unity of Science as a Working Hypothesis', in H.Feigl, M.Scriven, and G.Maxwell (eds.), *Minnesota Studies in the Philosophy of Science II*, University of Minnesota Press, Minneapolis, pp. 3-36.

3 Explanation

Natural science furnishes us with explanation, the human sciences with understanding. This dictum roughly captures the traditional *dualist* conception of the relation between the science of nature and that of human beings. The basis of this traditional distinction refers to the fact that choice of scientific method is determined by the nature of the ontological and epistemological realm of objects with which the particular science is concerned. Natural science uses the methods it does because it engages in the study of natural objects while the humanities have their respective methods because they investigate language and other forms of meaningful material. This distinction is linked to the idea that natural science avails itself of nomothetic and causal methods while the humanities are said to use idiographic and interpretive methods.

The motivation behind methodological dualism is, then, the fact that the natural sciences produce *causal explanations* while the human and social sciences yield, wholly or in part, *intentional explanations*. The dichotomy finds further motivation in the distinguishing feature of the human sciences: that they seek understanding through empathy or interpretation. It is, then, ruled out that interpretations are identical with explanations.

Not infrequently dualism is defended in a more radical form which goes so far as to assert that the human sciences, including in particular the interpretive sciences, neither *can* nor *should* concern themselves with the delivery of causal explanations. This approach is misconceived, however, since any account of an item's evolving impact over time must make reference to causal factors, and because any such account will tend to be reflected in interpretations of human conduct or art works. Even if that position were correct – assuming a difference between explanation in the natural sciences and interpretation in the human sciences – it is not obvious, however, how this might serve as an argument for the existence of a crucial difference in the use of methods.

As we shall see in what follows, there are good reasons for regarding interpretations in the interpretive sciences – that is, in the systematic study of literature, art, film, theatre and music – as explanations on a par with causal and intentional explanations in the natural and social sciences. The focus here is primarily on interpretations that seek to answer a why-question, and which we shall call *interpretive explanations*. We shall also see that in every form of intentional explanation and in every form of interpretive explanation a causal element is present. Consequently, it is never possible to point to a correct intentional or interpretive explanation without a postulated causal connection lying concealed within it. Yet, this makes neither intentional nor interpretive explanations identical with causal explanations. Science operates, as we shall see, with many different types of explanation that are irreducible to causal explanations.

The aim of science is to produce fresh knowledge about the objects of its research and investigation. Deeper insight and understanding so acquired yields explanations. Just as, in everyday life, we learn about things we do not understand by being given pertinent explanations so, by the same token, science seeks to deliver information about phenomena we are able to grasp only by being given a theoretical and systematic account (which may or may not include causal factors or causal accounts).

Hempel's explanatory model

Methodological dualism's continued dominance over many years traces in part back to Carl Hempel's classic attempt to give a definitive characterization of scientific explanation. For if Hempel's analysis is correct, scientific explanation is not the domain of idiographic and interpretive sciences. In his view, for any systematic account to qualify as a scientific explanation, it must fit into a specific type of explanatory model.

Hempel operates with two kinds of explanatory model: (i) the deductive-nomological model and (ii) the probabilistic statistical model. The idea here is that we have an explanation if (and only if) the phenomenon we seek to explain can be subsumed under a universal law or a statistical law. In the first case, the phenomenon in question (*explanandum*) is logically derivable (deducible) from the relevant law in conjunction with certain specified conditions (*explanans*). In the second case, we are unable to derive the event from the statistical law with logical necessity, but are able to predict the phenomenon's occurrence with a high degree of probability.

The explanation may be schematized as a logical inference:

The deductive-nomological model

Explanans	Laws Conditions	$L_1, L_2, \ldots L_n$ $B_1, B_2, \ldots B_m$

────────────── logically necessary

| Explanandum | Phenomena | $O_1, O_2, \ldots O_k$ |

The probabilistic-statistical model

Explanans	Laws Conditions	$P(O\|F)$ is very high F_1

══════════════ highly probable

| Explanandum | Phenomena | O_1 |

Hempel's conception of scientific explanation can summed up in the following points:

1. Explanation is an argument.
2. Explanation is thus a question of the logical derivation of a proposition about what is to be explained from a nomological proposition in conjunction with certain propositions about initial conditions.
3. Nomological propositions do not express physical necessity.
4. Nomological propositions need not refer to causal laws.
5. There is no logical difference between explanation and prediction.

Further, a number of conditions may be stated which any explanation must satisfy:

1. *Relevance*: the explanatory information must offer good grounds for the belief that the phenomenon actually exists/existed.
2. *Testability*: propositions that enter into an explanation must be empirically testable.
3. *Asymmetry*: the existence of the explanatory phenomenon is not itself explained by the phenomenon explained.

It is not difficult to see why Hempel's model is not conformable to the human sciences. An explanation only counts as a scientific explanation if the phenomenon can be subsumed under a general law. But most scholars in the humanities will reply that the humanities are not interested in laws because there are no universally valid principles in the human sphere. The disciplines within the humanities pursue *reflective* understanding of people, texts, works of art and musical works *qua* individual entities and seek to elucidate them in their wholly *unique* historical, social and cultural contexts. This, then, would seem to support the dualists' argument, but only if we assume Hempel's nomological model.

However, Hempel's explanatory models have met with vigorous criticism. First, many philosophers have pointed out that the model cannot be correct since in many cases it leads to paradoxes. Take, for instance, a flagpole and the shadow cast by it on a sunny afternoon. According to Hempel's deductive-nomological model we are able to offer an explanation of the length of the shadow on the basis of optical laws, given specifications of the height of the flagpole and the position of the sun in the sky. But on the basis of the very same laws we might explain the height of the flagpole, using our knowledge of the length of the shadow, and the position of the sun in the sky. This violates the asymmetry requirement. Consider another example: it is a statistical law that if an individual takes the contraceptive pill pregnancy will not occur. No men and few women have become pregnant while on the pill. So if Peter goes on the pill that fact will, according to the model, 'explain' why Peter does not become pregnant. But such an 'explanation' violates the relevance requirement. In other words Hempel's model guarantees neither asymmetry nor relevance.

Second, there seem to be many examples of explanations that cannot be subsumed under Hempel's deductive-nomological model. For it must surely be a reasonable demand to make of what purports to be a satisfactory account of the concept of explanation that it capture the ordinary explanations of everyday life as well as all types of scientific explanation.

Explanation as a rhetorical means of communication

Recent years have seen the development of a quite different conception of explanations, according to which they are rooted in the rhetorical practice of putting questions and giving answers. Bas van Fraassen's pragmatic theory of explanation, for instance, contains some of the basic ideas. It rests on four assumptions:

1. Explanation is not, as commonly assumed, a two-place relation between theory and phenomena but a three-place relation between theory, facts, and context.
2. Explanation in natural science is not importantly different from explanation in history or in everyday life.
3. Explanation is an answer to a why-question.
4. Questions and answers are not propositions with an unambiguous context-free meaning, but their content is dependent on the mode and the context in which they are uttered.

Point 4 may be illustrated through the same question being uttered in three different ways:

>Why did *Adam* eat the apple? – and not, say, Eve
>Why did Adam eat the *apple*? – and not, say, a banana
>Why did Adam *eat* the apple? – instead of, say, throwing it away

As the example shows, stress, intonation and communicative context impact on how the question is understood.

For an account to qualify as an explanation it is not necessary that all the conditions specified by van Fraassen be fulfilled. For instance, it is not clear why an explanation should be the only possible response to a why-question. As we shall soon see, answers to why-questions are merely one species of explanations.

Explanation is, I shall argue, a rhetorical practice, in the sense of being an intentional act of communication. 'Rhetoric', as the term is used here, has to do with expedient communication that is context-bound, directed and intentional, potentially persuasive, etc. An explanation is a response to a question from an interlocutor, and the explanation is intended by the respondent to furnish the questioner with the information he lacks – either by enlightening him about the facts, by making clear to him what the probabilities are, or by making abstract issues concrete. The respondent's answer is understood as an explanation because it places the information the questioner seeks into the broader context of what he already knows, or of what he is prepared to accept. Philosophers working on explanation usually focus their attention on the individual scientist who may himself raise a question. By making experiments and drawing on the results to articulate a causal story that accounts for the phenomenon he will offer an explanation. But by making this their focus those philosophers fail to take any account of the fact that the scientist is capable of raising such questions and answering them only in virtue of already belonging to a linguistic community. It is the latter that invests him with an understanding of what he is doing: of what it is to raise a question and to give an answer. What counts as an appropriate explanatory answer to an information-seeking question is determined by the public rules that govern speech acts involving more than one speaker. Therefore, explanation needs to be seen as part of a more general communicative practice.

Thus, for a fuller picture of what a reasonable account of explanation looks like, we should address the rhetorical features of this explanatory practice. It is important to recognize that explanation is an identifiable speech act that is successfully accomplished when it follows the unwritten rules governing the raising of an information-seeking question and the giving of an appropriate response to it. Explanation is, in other words, a matter involving far more diverse communicative rules and epistemic processes than merely those of logic.

First, explanation provides *understanding*. Making sense is what explanation is meant to do. It induces in us a psychological sense of knowing something – just as it very often put us in an epistemic state of actually knowing something. Again, philosophers discussing explanation tend not to concern themselves with understanding. It was basically this neglect that led to the tendency to want to see all explanation as having the logical structure of a formal argument. But facts about the world are not structured or arranged in the manner of premises and conclusions, so what good reasons do we have to claim that understanding, and by implication explanation, is properly delivered in terms of arguments? None! However, we do expect that explanation yields information that somehow enhances our grasp of the matter in question. If we know what is the case, we do not need explanation; the response does not add anything new to what we already know. If the interlocutor has no experience of coming to learn something new then, in his eyes, the respondent has failed to supply an explanation. The respondent must, indeed, have some idea of what she thinks would be an appropriate response before she can offer an explanation. But what counts as an explanation for her need not figure as such for the questioner. Only in those situations where her response fits with his background knowledge will her interlocutor be furnished with insight. Their understandings will be the same to the extent that they have a common epistemic background – which they do to a significant degree if they belong to the same linguistic community.

Second, explanation is *fact-oriented*. It makes reference to facts, or to what are at least taken to be facts. Information offered as an explanatory account is concerned with what is the case. But not all information about facts seems to count as an explanation. Factual information is necessary but not sufficient for explanation. If I for instance ask, 'Did Adam go to school today?' – and you inform me that he did by answering 'Yes' – could this response qualify as an explanation? It would appear not. An explanation does not merely consist in the citation of a fact; rather it tells us something about a fact by informing us about other facts. An explanation takes the form of a story that puts the information requested into a wider context.

Third, explanation is *truth-tracking*: One of our assumptions is that the epistemic value of explanation is not merely that it yields information about facts but, hopefully, information that is true. It is one of the questioner's objectives to obtain an explanation that is true and to have the respondent provide him with such. This does not indicate, however, that the force of explanation has anything to do with truth. We must, I believe, make a distinction between *force* and *value*. Many explanations are false although they still function as explanations. Aristotle's accounts of the movement of an arrow and his account of the fall of a stone are, despite being false, explanations nonetheless. If truth were essential for some account to have explanatory force, then much information offered by modern science as explanation would probably not qualify as explanation after all – in spite of the fact that we currently have good reasons to believe it. It is too strong a requirement that information must be true if it is to serve as explanation. A *correct* explanation is true, whereas an *incorrect* explanation is false.

Fourth, the explanans must be *relevant* to the explanandum: we must have good reasons to believe that the story being told is somehow connected to the fact being explained. Thus, a reference to the increased scarcity of storks in Denmark after the Second World War is not an appropriate response to the question why there was a sharp decline in the number of babies being born in the same period: these facts are simply not relevant to each other.

Fifth, explanations seem to be *asymmetrical* in the sense that the information explaining a fact is not itself explained by this same fact. The height of the flagpole together with the sun's position in the sky explains the length of the shadow. The length of the shadow, however, does not explain the sun's position in the sky nor the height of the flagpole.

Explanations are answers to questions that an individual puts because she lacks information about a specific situation or a specific phenomenon, and the answer is intended to supply her with the information she needs. Often the explanation-soliciting question is formulated as a why-question (to that extent van Fraassen is right) and the answer will accordingly consist of a reference to a connection between what we want to have explained and what we consider as delivering the explanation. It is also required that the answer be relevant. The required conditions for a response to qualify as relevant and informative in relation to a particular question are as follows: A relevant answer Q will consist in a statement of why the phenomenon P exists/occurs rather than one of a host of other possibilities X. And what is acceptable as a relevant answer Q is determined by our background knowledge.

Let us take a closer look at why-questions. It would be easy to imagine that all answers to why-questions express some type of causal explanation since their effects are precisely what causes explain. Reference to the cause explains why P occurs instead of being a total non-happening. But it is not always the case that the answer to a why-question states a cause. Of the following six examples of explanations only (a) constitutes a causal explanation:

(a) The pane was smashed because a stone was thrown at the window;
(b) This chunk of iron is rusting because it is a law of nature that iron oxidizes when in contact with water and oxygen;
(c) The elephant has a long trunk because of the selective advantage it offers in gaining access to leaves in the upper reaches of trees;
(d) The blood circulates in order to supply the various parts of the body with oxygen and nutrients;
(e) He went to the party in the evening in the hope of meeting her again.
(f) His antipathy to mice and rats stems from his repression of his childhood fear of his father.

Each of these six examples indicates a distinctive type of why-explanation that cannot be formulated analogously to any of the others. Each offers a particular explanation in virtue of its satisfying a formal structure which is peculiar to the explanation in question:

(a) *Causal explanation*: Such explanations appeal to the actual cause of a particular phenomenon, (the effect).
(b) *Nomic explanation*: Such explanations refer to the law governing a particular phenomenon.

(c) *Functionalist explanation*: Such explanations refer to the *actual* effect, *i.e.* a particular phenomenon is favourable or expedient for the reproduction or continuation of a particular individual.
(d) *Functional explanation*: Such explanations refer to the *actual* effect, *i.e.* a particular phenomenon is favourable or conducive to the survival or cohesion of the whole (be it a system or a society).
(e) *Intentional explanation*: Such explanations appeal to the *intended* effect of a particular phenomenon such as an action by referring to its aim.
(f) *Interpretive explanation*: Such explanations appeal to the *intended* effect of a particular phenomenon (be it a sign, a dream, an utterance, an action, a text, or a work of art) by regarding it as the expression of a particular symbolic, linguistic, literary or artistic meaning.

Out of these six types of why-explanations I shall here be concentrating on *causal, functional, intentional* and *interpretive* explanations.

Cause, function, motive and interpretation as explanation

The most prevalent type of explanation is, without doubt, *causal* explanation. It constitutes the typical response to a why-question and makes reference to the actual cause in explaining a particular phenomenon as the effect of precisely this cause. The causal explanation tells us why the phenomenon in question occurs (as the effect of the alleged cause) – and not some alternative phenomenon – by reference to the claim that the phenomenon in question occurs because the stated cause mediated it.

Indeed, many causal explanations occur as singular explanations; we explain one particular phenomenon by relating it causally to another: A person, Mary, contracted HIV because she had unsafe sex with a contaminated partner, Bill. But often a causal explanation takes the form of a nomic explanation involving a causal law. We must, nevertheless, not conflate these two kinds of explanation. They are conceptually distinct. On the one hand, we have various laws that are neither causal laws nor reducible to such laws. Only in physics can one point to symmetry laws and conservation laws. In case these laws figure in explanations we have nomic but non-causal explanations. On the other hand, it is debatable whether every causal fact entails the existence of a law. There seem to be many cases where a given fact is the cause of another – not because these facts are of certain types but because they have, in the particular context in question, the property of being the cause or the effect of the other.

Take, for instance, the extinction of the Great Auk. Mankind caused it, but what sort of law is involved in the explanation here? None it seems. Indeed, one could state that ecological conditions necessary for the survival of a species as deductive consequences of certain general biological principles, which a Hempelian wants to call laws. Nonetheless, even if causal facts do entail the existence of such a law, it would certainly not be in the sense in which laws traditionally have been understood, namely as universal relationships without exceptions. We shall return to this point in a later chapter. The main point is, however, not that there are no such laws involved but that they do not do the work of explaining. Obviously many possible perfectly contingent events could potentially cause the extinction of Great Auks; of them in this case we happen to know with great confidence that the one which actually occurred, and so caused the extinction of the bird, was the deed of men. So laws are not causes, men are.

Functionalist explanations are another type of explanations and quite different from causal explanations in the sense that they explain a phenomenon by pointing to what it causes instead of pointing to what causes it has. Some have sought to differentiate these different kinds of explanation on the basis of *content*, one type for 'inanimate' objects, the other type for 'animate' or at least for 'conscious' beings. We are so used to causal and nomic explanations in physics that we think that these are the only acceptable forms of explanation we have in this field. In biology, the social sciences or in the humanities, like linguistics, it may be different, but definitely not in physics. Many see the success of the physical sciences as resting on the fact that the philosophers and physicists at the end of the Middle Ages began to describe nature in terms of efficacious causes. I shall not deny this, but will argue that physics also uses functionalist or functional explanations. What I have in mind is the use of least action principles or the reference to conservation laws, states of equilibrium or to attractors in chaotic systems.

Let me give just one example. In atoms electrons can be in their ground state or in an excited state. If they happen to be in an excited state, they will sooner or later 'jump' back to their ground states (while radiating energy) because here they possess the least amount of energy possible. Moreover, physicists think that the 'jump' from an excited state to the ground state is an indeterministic process and therefore it is impossible to cite a cause of the event. Thus, we explain the movement of the electron from a higher energy state to a lower energy state by pointing to the effect, *i.e.* returning to the ground state. I shall call this kind of explanation a functional explanation.

The conclusion is that the rhetorical view on explanation leads one to hold that causal and functional types of explanation are not content specific but are used across the sciences. Since particle physics is, as it were, the least likely place one would find such teleological explanations, by showing they do occur there, we have effectively refuted the defence of retaining functional explanations as a separate content-sensitive types of explanation.

The difference between functionalist and functional explanations is parallel to the one between causal and nomic explanations. The first is concerned with individual phenomena. Nobody seems to doubt that we use functionalist explanation in biology. Here a property X (of a population) may be explained in virtue of the fact that X has better consequences for the reproduction of the individual organism than 'closer' alternatives to X. It is important to realize that the natural selection works on the level of individuals and not on the level of species or populations. As a consequence a property, which may be an advantage for the reproduction of a particular organism, may be a disadvantage of the population as such. The huge antlers of the Irish Elk are such an example. Stephen Jay Gould argues that the antlers *functioned* primarily as courtship signals to females. Therefore bigger antlers of a certain bull gave higher status and better access to the females, thus a better reproduction of that specific animal. But when the climate changed in Ireland, and the vast grassland, which was the natural habitat of the Irish Elk, was replaced first by the subarctic tundra and afterwards by heavy forestation at the end of the latest Ice Age, these enormous antlers became a disadvantage. The species became extinct because it failed to adapt rapidly enough to changing climate conditions or competition.

Also in the social sciences and humanities we find functionalist or functional explanations. In the social sciences this is in the form of functionalist explanation whenever somebody tries to give a biological explanation of human social behaviour. But in general this socio-biological approach cannot explain very complex social patterns because of the fast adaptation of human beings due to our intelligence. Social phenomena should rather be explained in terms of social causes or in terms of useful consequences. The American sociologist Robert Merton, for instance, has strongly argued in favour of functional explanations, maintaining that whenever social phenomena in relation to a group of people have beneficial consequences, intended or not, such phenomena can be explained in terms of these consequences.

Not everyone accepts the use of functional explanations in the social sciences. Jon Elster is very sceptical of their scientific value. Assume Z to be a group of people, X to be an institution or a behavioural pattern, and Y to be the function of X. We then say that X is functionally explained by Y if and only if:

1. Y is an effect of X;
2. Y is beneficial for Z;
3. Y is unintended by the agents who perform X;
4. Y – or at least the causal connection between X and Y – is unknown to the agents of Z; and
5. Y sustains X via a feedback mechanism through Z.

Although 4 may be challenged, Elster believes that what is most problematic is that 5 is taken for granted rather than being demonstrated. He builds his objection on two premises: (i) functional explanations are successful only if we have reasons to believe in a feedback mechanism from the consequence Y to the phenomena X to be explained, and (ii) such reasons for believing in a feedback mechanism must and can only be demonstrated in each and every single case. Condition (ii), however, does not hold in biology, Elster argues, because the theory of natural selection gives us a general knowledge about such a mechanism. So the question is how fatal Elster's objection is.

First, it seems correct to say that we do not need to know the existence of a feedback mechanism in order to give a functionalist explanation. On the rhetorical view on explanation, functionalist accounts get their explanatory power from the fact that such answers provide us with new information about a phenomenon X by saying that it continues to exist because it has a beneficial effect Y for Z. This information may suffice as an explanation depending on the context and the interlocutor's interest. Nonetheless, the interlocutor may ask for further information about the underlying mechanism that allows Y to maintain X, and then the question of causal feedback processes becomes quite relevant.

Second, even if sociology cannot come up with a general idea of such a mechanism like the notion of natural selection, the more individual account is no less explanatory than the nomic account, it is merely more idiographically focused. However, it has been argued that there exists a general notion of a mechanism that can be used to explain the beneficence (or the sustenance) due to the idea that social systems have a presumption for equilibrium. Thus, Arthur Stinchcombe suggests that a social change can be looked at as an absorbing Markov chain. But such an idea does not escape the introduction of the functional account; it does not provide us with a causal feedback process. Pointing to the fact that social systems have a tendency to evolve into a state of equilibrium and remain there is an explanation in terms of its effect. But this is quite similar to explanations in physics that appeal to states of equilibrium, conservation, attractors, or ground states. Sometimes we have to accept as a brute fact that we cannot come up with a causal explanation of nature's preference of such states, but we can use these 'preferences' functionally to explain other things.

A third group of explanations comprises those given by reference to motive and called *intentional* explanations. These are responses to why-questions in which reference is made to the intended outcome in order to explain the particular action. The intentional explanation tells us why an agent performed one action and not another by indicating that the action was chosen because the agent sought to realize a particular end and judged the action to be an effective means towards achieving that end.

Explanations in terms of motives have been the subject of much discussion, for it has been contended that motives cannot be causes. The philosopher Gilbert Ryle, for example, has argued that motives are dispositions and cannot therefore figure in causal explanations. To illustrate: the glass broke (i) because it struck the edge of the table, or (ii) because it was brittle. The first is a causal explanation, the second a law-like judgement, says Ryle. The idea behind this distinction appears to be that the situation in which the glass strikes the edge of the table is an event which occurs at a particular place and at a particular time, whereas a disposition is extended in time and space, and figures as a standing condition in many places and at many times.

Against Ryle's claim, Elizabeth Anscombe has rightly objected that there are many motives that may occur just once in a person's life without their figuring as dispositions. If a person tells an untruth on just one occasion, this does not mean that she is mendacious (has a disposition to lie). It would, in consequence, be more correct to say that psychological dispositions are not motives precisely because actions that spring from dispositions cannot be intended. An individual may be quick-tempered without any desire to be so.

There is, however, a much stronger argument available to support the claim that motives are not causes. Abraham Melden has advanced the so-called logical dependence argument. Cause and effect are logically independent of each other: neither is describable in terms of the other. A volcanic eruption may be the cause of a town being razed to the ground. But we can describe a volcanic eruption in terms that make no reference at all to the destruction of towns, just as we can describe the destruction of a town without volcanic eruptions figuring in it. Consequently, then, causal connections are not logical but contingent connections between independent events, and are describable in terms of *synthetic* propositions.

Motives and actions are, by contrast, not logically independent. A motive cannot be identified without reference to an action. If we describe an action as a murder, the motive to murder someone is implicit in our description of the action; or if we describe an action as being that of going to work we refer to the motive of going to work in our characterization of that behaviour. In other words, motives and actions are logically connected. The relation between motives and actions, says Melden, can therefore only be expressed in *analytic* sentences and this means that motives to action are excluded from the category of causes of action.

It is not easy to demolish the logical dependence argument. Donald Davidson, for example, has asserted that synthetic judgements may be converted into analytical ones, and vice versa: '*A* caused *B*' may be written as 'The cause of *B* caused *B*'. But this objection does not seem to deliver a satisfactory answer to the problem. Where Davidson is right is in arguing that explanations in terms of motives have a causal element attached to them even if they cannot be reduced to causal explanations. Despite it being true that we describe an action by reference to a motive, a given description may not reflect the actual motive. But of course the correct explanation would. As we come to learn more about an action we may discover that its motive was other than that suggested by its description. An agent may well perform an action ostensibly describable as the quenching of his thirst, whereas in actual fact he may just be simulating that action, in the knowledge that he is being spied on by another.

The fact that explanations are configured differently in terms of motives than causal explanations is critical to when we wish to characterize actions in light of their motives. Nevertheless, the only plausible understanding of motives seems to be that they function as the causes of the corresponding actions. How is the dilemma to be resolved?

Part of the answer is to be found in an argument of Georg Henrik von Wright. His answer rests on a methodological observation rather than a properly logical account of our concept of motive. He claims that the logical dependence argument draws its force not so much from the fact that we cannot refer to the motive as that we cannot *verify* the motive without reference to the action. Conversely, we can only understand an action if we can connect it to a motive. It is these features, I shall argue, that the criterial account uses in determining the meaning of motives.

The criterial identification of a concept differs from its logical counterpart: according to the former not all concepts are explicable in terms of the necessary and sufficient conditions for the truth of a sentence. It is customary to say that a concept can be explained by reference to the conditions determining whether the sentence containing the concept is true or false. But in certain cases we cannot explain the meaning of the sentence without using criteria.

Conceptual criteria

(a) The connection between criteria and what they are criteria for is *evidential*, *i.e.* the criteria are, logically speaking, neither necessary nor sufficient. The connection is, in other words, contingent.
(b) The fact referred to in (a), namely that the relation is evidential, must be a logical fact. This means that it is part of the *meaning* of the word introduced or explained that the criteria constitute good evidence for the word's being used correctly. The connection between the meaning of the word and the criterion is a logical relation.
(c) The criteria are defeasible, *i.e.* they can be rejected because (a) states a contingent relation.

To understand the difference between (a) and (b) we must note that (a) speaks of a connection between two non-linguistic entities while (b) concerns a connection between a linguistic expression and the criterion for the expression's use. There are a host of examples of relations that satisfy (a) but not (b), or vice versa. The barometric pressure of 980 millibars is evidence of a storm, but that a storm makes the barometer show 980 millibars is not part of the meaning of the word 'storm'. There is merely a causal relationship between them. By contrast, 'a bachelor' will always denote an unmarried man. There is, then, a logical connection between being a bachelor and being an unmarried man. For if we know that a man is a bachelor, we know automatically that he is an unmarried man because 'unmarried man' defines the meaning of the word 'bachelor'. Neither of these examples satisfy both (a) and (b).

If, by contrast, we apply the three conditions to motives, they would seem to accommodate our various intuitions. The connection between motive and action is not expressible as an analytical relation, but it is part of our concept of a motive that the action putatively springing from it constitutes good evidence in support of our identification of it. We can sum this up in the following requirement:

The criterion for a motive being regarded as an intentional cause is the evidence we have for the motive, because it is part of the meaning of the word that there is good evidence for its being used correctly. The evidence referred to is behaviour which does not just follow from the motive but, *qua* evidence, must be caused by it.

For instance it is consonant with the concept of vindictiveness that a rejected lover, eager to take revenge on his rival, will aim at doing something drastic to harm him. Let us imagine that I see Alan set fire to James's house, and I know that James now goes out with Olga who was formerly Alan's girlfriend. We might then say that his setting the house on fire is a sign of Alan's thirst for revenge because the use of the term 'thirst for revenge' is such that Alan's behaviour may be taken as a sign of the presence of that motive. On the other hand the house being set alight is only an indication of Alan's vindictiveness, because he could be setting fire to James's house for other reasons than those suggested here.

We can now use this motivational analysis to characterize intentional explanations. The motive behind an action is to direct the action towards a particular end. An intentional explanation is one aiming to show that the behaviour that is to be explained (*explanandum*) is meaningful in relation to the desire purported to explain the action (*explanans*) because the behaviour is both evidence for, and can only be understood as meaningful by reference to, the motive. There is thus a contentual relation between *explanandum* and *explanans* that emerges through the action's propositional content being described in terms of the propositional attitude (the desire). However, behaviour can only figure as evidence for a motive and be grasped as meaningful, if the following three requirements are met:

(i) The person himself must regard the behaviour in question as a means, or as the best means, of realizing his desires.
(ii) The person's desires and beliefs must be the cause of the behaviour.
(iii) The desire and the beliefs must cause the behaviour 'in the right way'.

There are good arguments for the claim that these conditions must be satisfied if an explanation is to count as a correct intentional explanation. The expression 'in the right way' indicates that the action produced by a particular desire must conduce to the intended goal. A marksman who misses the mark through sheer nervousness at the thought that a bull's eye would make him Olympic Champion patently fails to score a bull even though that was what he wanted to do. In other words behaviour must not be accidentally linked to motive and aim. The resultant effect must also be the intended effect. Something counts as an intentional explanation if and only if the following conditions are satisfied: An action is, in the view of the agent, a means (and possibly the best means) by which to realize her wishes. These desires and beliefs in their roles as reasons render the behaviour intelligible, but the conformity between desires and beliefs and the action itself must not be due to some contingency. The action must be performed because it accords with the agent's desires and beliefs.

In addition to causal and intentional explanations we have *interpretive* explanations. Such explanations are a response to why-questions in contexts where what needs explaining is the occurrence of a sign, an action, a text or a work of art, through its identification as the expression of a particular meaning. For example, the choice of a text with a particular literal meaning rather than one with a different literal meaning is explained by reference to particular figures, symbols, narratives, etc. for which it is the vehicle. An interpretive explanation tells us, then, why the text is configured as it is by pointing out that this particular choice is the most effective means of expressing the symbolic, metaphorical or literary meaning evinced by the resultant text. We understand (the choice of) text by adverting to its symbolic and metaphorical meaning.

An interpretive explanation in literary studies sees the work's *direct* or *literal* meaning as a means, possibly the best means, of conveying the narrative intentions as these come to expression in the work's *symbolic* or *metaphorical* meaning. Indeed one might say that, really, the relation between authorial intention and the text's symbolic or metaphorical meanings constitutes the intentional element in the explanation, while the relation between the text's immediate, literal meaning and its symbolic meaning is the true interpretive element in the explanation. However, to avoid unnecessary complications we shall refer to these grounds taken together as the interpretive explanation.

Interpretive explanations do not exhaust the interpretive enterprise: interpretive explanations only occur as answer to why-questions when the enquiry concerns the choice of a sign, an action, a gesture, an utterance, a text, a dream, etc. Interpretation so understood invokes the literal meaning of the work in stating what is true, say, of the work's narrative intentions. Every reading in interpretive science seeks an explanation of the meaning of the choice of a particular sign, action, text, work of art, etc., and seeks to know the writer's or artist's reasons for expressing herself as she did – grounds attaching to the sign, action, text, work of art as the bearer of representational meaning. There seems here to be a causal connection between that of which the *explanans* (that which explains) speaks and what the *explanandum* (what is explained) refers to. It is natural to think that authorial intentions are the cause of the text's symbolic content, which in turn is the cause of the choice of a text with a particular literal meaning.

The problem is acute for psychoanalysis. For many years a vigorous debate has raged as to whether psychoanalysis should be understood as an empirical theory or whether it can and should be understood as a hermeneutical discipline. Psychoanalysis explains dream or neurotic behaviour through the claim that an unpleasant experience that has been repressed in the subconscious later eventuates in a dream or a neurotic action bearing some thematic affinity with the original experience. Paul Ricoeur has emerged as the proponent of the hermeneutical reading whereas Adolf Grünbaum argues for its empirical counterpart. Ricoeur is right when he says that psychoanalytic explanations are neither causal nor intentional, but he reaches this conclusion on the basis of erroneous premises. He claims that such psychoanalytic statements cannot refer to causes since motives (mental events) cannot function as causes. Nor can they refer to the actual motives, of course, for these are subconscious. Rather, they bear the semblance of causal explanations without being such. Grünbaum is right to draw attention to the fact that thematic affinity between mental events cannot serve as the criterion of a causal connection between them. Relatedness with respect to content is not the same as a causal connection.

Evidence going beyond a thematic relation is required to demonstrate the presence of a casual connection. But he is mistaken in saying that hermeneuticists confuse the meaning of natural signs with semantic meaning. The hermeneuticist can agree if the doctor says to the mother that the red spots on her child's skin are a sign of measles, where the 'meaning' of the symptoms is determined by their cause, this is not the same as saying that that the symptoms in a semantic sense designate the cause. Grünbaum thus draws attention to an important problem. For if neuroses and dreams *must* be understood as the result of causal processes in the unconscious, how can neuroses and dreams have symbolic meaning which does not simply rest on a natural semiotic relation between cause (object) and effect (sign)?

The answer is simple. In every interpretation that seeks to explain a neurotic action or a dream *qua* representation of unconscious intentions, a causal element is present. But this does not mean that either neurotic acts or dreams are thus turned into natural signs because *qua* signs they are intended, albeit unconsciously, to represent a repression. It is intention that marks out the sign as either natural or semantic.

If we compare the respective structures of causal, intentional and interpretive explanations, we see that in what concerns causal explanations only a causal element is present. An indication of the causal connection between two phenomena is a necessary and sufficient condition of the existence of a causal explanation. In the second type of explanation, intentional explanation, both a causal and an intentional element are present. Each of these two elements constitutes a necessary condition for the explanation's qualifying as an intentional explanation. And in the case of interpretive explanations, the third kind of explanation, interpretation, causal, intentional, and interpretive elements are all present and are each necessary conditions for qualifying as an interpretive explanation.

In other words, we may conclude that it is not so much the absence of causal explanations as the presence of intentional and interpretive explanations that points up the crucial ontological difference between the natural sciences and the humanities. But, and this is the next step, even if we recognize that intentionality and meaning are essential features of human persons and human activity, this still does not mean that explanations delivered in terms of these features are the result of a particular scientific method. The traditional conception still needs to show that causal explanations as well as intentional and interpretive explanations constitute distinctive *methods*. In my view this cannot be done because explanation and method are logically disparate categories.

Epistemic goals

There are other kinds of explanation than those that answer why-questions. If explanations are understood in terms of their rhetorical functions in a communicative practice, it is difficult to retain a conception of explanations as answers to why-questions as van Fraassen suggests. Utterances delivered as responses to what-, how-, in what manner-, when-, where to-questions, also function as informative answers if we are in a situation where we lack information and the information solicited is given us by reference to facts other than those to which the question was directed. Critical to the form the question takes is the issue of the *epistemic goals* the enquirer expects the answer to meet.

If, for instance, I ask *why* the pyramids were built, the answer that they were built as tombs for the Pharaohs counts as an explanation. What is offered, then, is an intentional explanation. But if I ask instead *how* the pyramids were built, I am asking something else and the answer that they were built as tombs for the Pharaohs will not count as an explanation. By contrast, a story telling us that the Pharaohs used thousands of workers, and that the Egyptian builders had primitive hoisting apparatus and ramps at their disposal will be regarded as an explanation. Such an answer offers what might be called an explanation in terms of *form*. There are other types to set alongside it, such as *material* and *structural* explanations, which, similarly, may not figure as answers to why-questions. We might have asked *what* the pyramids are made of, *how* they were constructed and satisfactory answers to these questions would make reference to the composition of the materials and the geometrical and spatial design of the structures. To the question of *when* the pyramids were built, an answer to the effect that they were erected between 2000 and 4000 B.C. does not, in itself, serve as an explanation. Simply to state a time, the expression of a fact, is not to offer an explanation. An answer counts as an explanation if it refers to a fact already known (such as that consisting in the dating of the pyramids) and grounds that fact. So if it is advanced that the inscriptions on the walls indicate that that they were built as monuments to certain named Pharaohs, and we know from other sources when these kings lived, then we have given an explanation in response to a when-question. Explanations are answers or accounts which instruct us about something not previously known to us by setting the fact focused upon in relation to other facts, and the speaker's and the listener's relevant background knowledge. In any event the specification of relevance given above in relation to why-questions also applies to answers that are not responses to why-questions. However, this extension of the concept of explanation will not concern us further here. Let it simply be added that responses to some of the types of question other than why-questions may also involve interpretation, as exemplified by the question: 'What does X mean?' where X's meaning is given by a general linguistic rule. An interpretive explanation will not serve as an answer to this question.

We have ascertained, then, that why-questions are appropriately met by generically various answers. We still need to show, however, what it is that leads us to look for one sort of response rather than another. The answer is similarly straightforward. The response we find satisfying in answer to a given why-question, *i.e.* whether it is a causal, intentional or an interpretive answer we want to hear, is contingent upon our *epistemic goals*. And our epistemic goals are of course determined by what we hold to be true about the domain in question. If we believe it to be a fundamental feature of nature that phenomena are causally connected, it will naturally be an epistemic goal of natural science to answer why-questions by reference to a cause, since it is the goal of explanations to be true. By the same token, it might be said that if we believe that it is a fundamental feature of human beings that their conduct is informed by intentionality it naturally becomes an epistemic goal of the social and human sciences to offer answers to why-questions with reference to the beliefs, motives or desires underlying human behaviour and linguistic acts. In such cases we shall make the intentional explanation our epistemic goal, since such a description tells us what holds true in the human sphere. And lastly we may say that if we regard meaning as a basic feature of the products of this behaviour, it is naturally an epistemic goal to answer why-questions about a work of art by reference to an interpretation that explains the meaning of the work, because we want what the interpretation says about it to be true.

Truth is normally regarded as the highest epistemic goal even though it has been argued that our epistemic goal is simply empirical adequacy, *i.e.* that our theories need merely to be in agreement with our observations and experiments. But regardless of which epistemic goals we formulate, these are not in themselves methods. Methods are rather what aid us in realizing such goals. There are, accordingly, good reasons for separating epistemology and methodology since epistemic goals can only be met systematically if the requirement imposed on our methods is that they be reliable. Whether or not they are reliable is a factual matter whereas choice of aims is a normative one.

We have until now accepted that the difference between inanimate nature and human beings is that, in contrast to physical processes, human acts are meaningful and intentional. Against that background, we have argued that both causality and intentionality constitute important features of the world, features to which we refer in those of our explanations that are offered in response to why-questions, because we want such explanations to be true. We have rejected, then, the clause that causal explanations and intentional explanations constitute different methods relative to the natural and human sciences, respectively. But by so doing we have naturally not yet demonstrated that there is no need for methods in the humanities beyond those used in the natural sciences.

It is, however, difficult to see what the arguments for a logical connection between an ontological domain and general choice of method should be. Naturally, it depends to some extent on how we define scientific method. But if we say that a scientific method is a procedure that enables us to determine whether a particular belief about the world is true or false, there is nothing in this characterization of scientific method that requires that different methods be selected relative to the species of truth we are after.

Methodologists distinguish between methods of discovery and methods of justification, even if there are grounds for claiming that justificatory methods are simply particular procedures aimed at discovering whether a hypothesis is true or false. The method of discovery functions by taking us from the observation of facts to a hypothesis that purports to explain these facts, whereupon the method of justification is introduced to determine whether the hypothesis in question is true or false. In the chapters that follow I shall be examining whether the basic concepts defining this process of discovery are as appropriately and successfully applied in the humanities as they are in the natural sciences.

References

Anscombe, G.E.M. (1957), *Intention*, Oxford.
Cohen, G.A. (1978), *Karl Marx's Theory of History. A Defense*. Princeton University Press, Princeton.
Davidson D. (1980), 'Actions, Reasons and Causes', in *Essays on Actions and Events (1962)*, Oxford.
Elster, J. (1983), *Explaining Technical Change*, Cambridge University Press, Cambridge.
Faye, J. (1999), 'Explanation Explained', in *Synthese*, 120, pp. 61-75.
Gould, S.J. (1974), *Ever Since Darwin. Reflections in Natural History*, Penguin Books.

Grünbaum, A. (1984), *The Foundations of Psychoanalysis: A Philosophical Critique*, University of California Press, Berkeley.
Grünbaum, A. (1988), 'Précis of The Foundations of Psychoanalysis: A Philosophical Critique', in Clark P. & Wright C. (eds.), *Mind, Psychoanalysis and Science*, Blackwell, Oxford, pp. 3-32.
Hempel, Carl G. (1964), *Aspects of Scientific Explanation*, The Free Press, New York.
Melden, A.I. (1961), *Free Action*, London.
Ricoeur, P. (1970), *Freud*, Yale University Press, New Haven.
Ryle, G. (1949), *The Concept of Mind*, Middlesex.
Salmon, W. (1989), 'Four Decades of Scientific Explanation', in Ph. Kitcher & W.C. Salmon (eds.), *Scientific Explanation, Minnesota Studies in the Philosophy of Science*, vol. XIII, University of Minnesota Press, Minneapolis.
van Fraassen, B. (1980), *The Scientific Image*, Claredon Press, Oxford, Chap. 5.
von Wright, G.H. (1971), *Explanation and Understanding*, London.

4 Interpretation

Natural sciences explain, human sciences understand. Thus ran the proclamation of the German historian Gustav Droysen in 1858 and, thanks to the philosophical writings of Wilhelm Dilthey, it has been a credal tenet of scholars ever since. Understanding in the humanities requires empathy and sympathetic insight. Today that message is inextricably bound up with the assignment of the hypothetico-deductive method to its proper use in natural science, and of the hermeneutical method to the humanities and the interpretive sectors of the social sciences. We shall return to those methods in a later chapter. Here I shall show that natural science seeks understanding to the same extent as do the humanities, and applies interpretation in its efforts to arrive at such understanding.

Indeed, if we define understanding as grasp of meaning, the definition would rule it out that the natural sciences being capable of producing any sort of understanding. If that were the case Droysen's dictum would only be stating the obvious, namely that the sciences of nature deal with objects quite different from those dealt with by the sciences of man. Although this is the received view, it is a misguided attempt, based on linguistic definitions and the conflation of ontology and epistemology, to say something profound about the acquisition of knowledge. The notions of explanation and understanding are epistemic in the sense that they are concerned with the cognitive acts of gaining and handling knowledge.

We saw in the preceding chapter that scientific explanations are not confined to the natural sciences. Explanations figure to an equal extent in the human sciences. Researchers and scholars seek explanations when faced with things they do not understand and a good explanation remedies this deficiency. For example, when palaeontologists make the discovery that dinosaurs became extinct around 60 million years ago they had no explanation as to why it happened. According to *one* explanation their extinction occurred because a meteor hit the earth and vast quantities of material were hurled up into the air. This led to the formation of a dense cloud cover that the sun's rays were unable to penetrate; the temperature dropped and the natural sources of dinosaurian sustenance dried up. This explanation enabled palaeontologists to make sense of their discovery: they could now deliver a story as to how dinosaurs became extinct – a story which put their findings in a concrete context.

The same applies when it comes to making sense of events in everyday life. If I discover brown patches on the lawn, I only understand how they got there when I learn that my wife has been out spraying weed-killer. Once able to put the patches into context I gain an understanding, which I had lacked until that point. The explanation gives me an insight into why the patches are there.

Understanding is partly the psychological insight one acquires on getting some sort of fix on the nature of things, and partly the epistemic grasp enabling one to explain, say, why they are as they are, or how they are as they are. There are philosophers of science who hold that, as compared with classical physics, quantum mechanics fails to offer understanding because we can neither envisage what takes place in the atomic world nor explain it. How can something be both a wave and a particle? A causal explanation, by contrast, engenders understanding. Nature becomes meaningful to us when, by analogy with what we already understand, we are able to get a grip on what had previously eluded us. What occurs in nature can be made meaningful to us without nature itself needing to be meaningful.

Understanding is not the exclusive province of scholars. But the domain of objects to be explained in natural science is normally distinguishable from that in the human sciences. The important point would seem to be that while all kinds of explanation yield understanding, some explanations have understanding as their subject matter. In such cases the explanations are other than causal explanations. There is a difference between simply *grasping* the fact that a particular causal nexus holds (and thereby making it meaningful to us) and *grasping* the import of something that is already meaningful. The difference has to do with the ontological constitution of the item concerned and, correspondingly, choice of type of explanation.

Interpretation in the human sciences

Everything can be an object for the understanding but only a minority of things embodies it intrinsically. Few things are meaningful objects. The category of meaningful objects includes such things as words, pictures, texts, acts and persons. Such are the items that are the materials of the humanities. So it is, the argument goes, that the humanities seem to have understanding of meaning as their aim. For human beings have the facility of self-expression, and are able to express themselves through writing, painting, dancing, singing and the delivery of a clip on the ear. The result is a text, a painting, a song or a thick ear – expressive of that to which the individual sought to give expression: a thought, feeling, mood or attitude. What is expressed is also called meaning, and everything that expresses meaning in that way is termed meaningful. Accordingly, acts, texts and paintings are meaningful because they represent that for which expression is sought.

Meaning may be immediately accessible. If I observe a woman cyclist stop at a red light I understand right away that she is simply complying with the Highway Code, which dictates that no vehicle may pass through a red light. I do not need to interpret her behaviour in order to see the action as meaningful. I see the action as expressive of her compliance with a rule. Here meaningfulness is an inbuilt constituent of my perception of the action. I do not witness a particular sequence of physical behaviour that I then set about interpreting in order to discover the meaning. The meaning is straightforwardly obvious to me in the sense that my understanding of it is *non-inferentially* acquired. So it is too if I observe a lion. I recognize immediately that it is a lion the very instant I set eyes on it, because I know what lions are, and that the word 'lion' applies to the specimen currently before me. I do not begin by experiencing sense data which I then read as indicative of the presence of a lion. The concept of a lion forms part of my perception of lions, just as the concept of stopping at red lights enters into my perception of the cyclist's behaviour. So, we also use intentional explanations in cases where we have no difficulty in understanding how what we seek to explain is to be understood.

But if I do not immediately understand an action I seem to have no option but to resort to interpretation. Even if acts, texts and paintings are meaningful, it is not always the case that the meaning they express is perspicuous to the viewer. Consequently, he or she must produce a construction or an interpretation that uncovers the meaning initially hidden from him or her. It would seem to be precisely this bringing to light of hidden meanings not immediately accessible that is the task of the human and social sciences. So it becomes an important feature of these sciences that they produce interpretation, *i.e.* the satisfaction required for understanding.

In an influential article, 'Interpretation and the Sciences of Man', Charles Taylor draws attention to the fact that interpretation is essential to explanation within the human sciences, that interpretation consists in the elucidation of the object of one's enquiry, and that such elucidation is hermeneutics. Unfortunately, it quite escapes his notice that natural science is rich in interpretation. And the reason for this oversight is simple: Taylor mistakenly believes that natural science builds on what he calls 'brute data', *i.e.* raw facts that go uninterpreted, while social and human sciences operate with facts that can only be understood *qua* interpreted. We shall return to this rationale, but first let us summarize his analysis of elucidation.

Taylor sets out three conditions for interpretation:

1. There are objects that have meaning or significance attached to them.
2. Distinctions may be drawn between context and absence of context, between meaning and the bearers of meaning, between meaning and its expression, even if such distinctions are relative.
3. There must be a subject capable of setting out criteria for similarity and difference, for otherwise the choice between these criteria would be arbitrary.

Meaning is not free-floating. It attaches to physical objects such as sounds, signs and behaviour. These are bearers of meaning when they enter into contexts with other meaning-bearing sounds, signs and behaviour. A sign that does not enter into a system of meaning-bearing signs cannot itself be a bearer of meaning. We can nonetheless distinguish between signs and their meaning, between expression and content. The same meaning can be rendered by a variety of expressions. Consider the sentences 'Snow is white' and 'La neige est blanche' – two different expressions with the same meaning. Further, there must be a person, a subject capable of grasping the context that confers meaning on the individual object.

Taylor argues that there is a difference between linguistic meaning and what he calls experiential meaning, but holds that they have three features in common:

1. Meaning is meaning for a subject.
2. Meaning is about something; we can distinguish between a given element (situation, action, proposition) and the meaning it has, and so we can describe the element in two ways.
3. An element can only have meaning in relation to a semantic field (semantic holism), in which the element's meaning stands in contrast to

the meaning of other elements. There are no elements with a meaning that can be isolated from other meanings.

Experienced meaning is for a subject, is about something, and occurs in a field, whereas linguistic meaning in addition concerns the world of referents. Here:

4. Meaning is the meaning of signifiers.

The definition of a word is a statement of the meaning the word has for us. But when Taylor talks about elements with experiential meaning he has in mind such items as colours, feelings, attitudes, desires and acts.

Critical to interpretation is our ability to distinguish between the element and its meaning. Were the element not distinct from its meaning there would be nothing to interpret. The bare recognition of the element would immediately yield its meaning. But that would require that element and meaning were analytically connected whereas, in fact, their correlation is causal or conventional. So we can either describe the element as a purely physical expression, or as meaning-bearing – just as we might describe a person's conduct in terms of physical behaviour or of meaningful agency.

Taylor also poses the question as to how we determine whether an interpretation is correct. What decides this, he says, is whether the interpretation creates clarity and connectedness in what we want to understand. An action can be regarded as meaningful if, from the point of view of the agent, there is a connection between his action and the import of the entire situation in which he finds himself. But plainly a difficulty arises if meaning can only be described in relation to other meanings. For if the meaning of A explains the meaning of B, whose meaning in turn explains C, which, by the same token, explains A's meaning, we seem stuck with a classic case of circular reasoning. How do we avoid interpretation becoming circular?

The solutions offered by both rationalism and empiricism are unsatisfactory. Taylor stresses the impossibility of escaping the hermeneutic circle, mediated by the fact that individual elements are interpreted in the light of the whole, which in turn is to be understood in the light of the individual elements. The meaning of a word is explained against the background of other words, a text in the light of other texts, but the meaning of an emotion or action is not explained by reference to other emotions or acts but, rather, in accordance with their linguistic specification. The relation between verbal meaning and experiential meaning is not that of feelings, desires or needs being natural kinds existing independently of how they are described, nor that of thought producing feelings, desires or needs regardless of whether or not they are psychologically possible. It includes, says Taylor, something of both. Human beings are self-interpreting animals. Meaning and interpretation are interwoven. What is interpreted is itself an interpretation. There exists no meaning independent of our interpretation of that meaning. It is the interpretation of experiential meaning that lays down the meaning of that meaning.

In my view Taylor's analysis is untenable on three important heads: His understanding of natural science is somewhat outmoded, he exaggerates the role of interpretation in the humanities, and his restriction of interpretation to meaningful material has no warrant in reality. To begin with, there are no immediate data in natural science to be set against interpreted data in the humanities. Our data are dependent on our theories, background knowledge, and linguistic competence, in both areas. Our concepts of the world enter into our experience of it. I do not *interpret* nature, human beings or texts when I look at the flowers in my garden, attend a church wedding or read the paper. I have recourse to explanations if I have to say why something happens or what it means but I do not have to interpret what I see or read in order to understand it. Experiential meaning is not the result of an interpretation. It is part and parcel of our perception of the world, and of the linguistic and social practices attached to our seeing the world as we do instead of experiencing it in some altogether different way. We need an interpretation only when we fail to understand the meaning of what we see, hear or read. Interpretation provides understanding but understanding occurs to a vast extent without any interpretation.

In my view it is equally mistaken to seek to tie interpretation to empathy or sympathetic identification. The fact that interpretation in the humanities often makes reference to emotions, desires and actions does not mean that some special affective experience is required to understand them. We can often identify with something we immediately understand. If I hear that a good friend is seriously injured, I do not have to envisage that situation in vivid detail in order to interpret it – rather, I feel for him in that situation because I already understand it. On other occasions I interpret a person's motives without identifying with them. An interpretation of the motives of Milosevic's henchmen will hardly, for most people, be accompanied by a strong sense of identification with their mindset. So even though feelings are in many cases bound up with interpretation these are mutually independent. Empathy and sympathetic identification can motivate (and arise from) an interpretation without being a part of it. Interpretation conducive to scientific understanding can involve empathy and sympathetic identification, but must always satisfy the epistemic standards that are the marks of rationality.

But when do we have an interpretation, then, and what counts as an interpretation? If interpretation yields understanding it may be seen as a process productive of understanding on the part of the interpreter. But in that respect interpretations do not differ significantly from explanations. Every explanation can be seen as a process which mediates understanding in the questioner. However, interpretation too can be the result of such a process – which is to say that it can be the *tentative* explanation of something that was previously not understood. Therefore we have as many kinds of interpretations as we have kinds of explanations, but only one kind can be associated with interpretive explanation. What is distinctive in the humanities about the use of intentional and interpretive explanations, is that what is interpreted is experienced as representing something other than itself. One thing stands for another. It is not enough that a thing be (causally) connected with another thing. The connection between the two must be intended if the one thing may plausibly be said to represent the other. It is a constraint on any interpretation that what is interpreted extends beyond itself and refers to something other than itself. Every interpretation consists, as it were, in a construal of what this other thing is, and the interpretive explanation, for instance, offers an answer as to what it is.

In light of this we can state a provisional definition of interpretation:

(I) The connection between X and Y constitutes an interpretation for a person P, if and only if (i) P believes that X represents Y, because X is intentionally connected to Y, and (ii) P's belief is the result of a hypothesis.

An interpretation is based on an assumption about what X stands for, and an interpretive explanation gives us an understanding of the meaning of X by framing an adequate hypothesis about X's connection with Y. Accordingly, it is possible for X to represent Y, while it is part of the understanding of X that X is intentionally connected to Y without that understanding resting on an interpretation.

For instance, I understand the meaning of the word 'rose' and, as implicit in this understanding, the fact that the word stands for a rose. My competence as a language user – *in casu* English – is evidenced by my immediate understanding of what the word refers to, without my first having to embark upon a process of interpretation through which the word is associated with the object. This understanding is practical knowledge deriving from my command of English. It is not part of a theoretical hypothesis which merely expresses a probability. But if the word 'rose' appears on the page of a book it may not just stand for rose in the literal sense but may also, *qua* metaphor, symbolize true and everlasting love. That the latter meaning is intended will normally only emerge through an interpretation.

The same distinction applies in the case of actions. Often we are able to understand an action performed by an individual without having to interpret his behaviour, but in other cases we can only understand it in the light of an interpretation. When a man holds the door for a woman he is with, she normally has no difficulty in understanding that he is inviting her to go first. Her understanding rests upon the fact that she inhabits a culture in which it is usual for men to hold the door for the women they are with. She does not need a hypothesis about his motive to guide her in understanding his conduct. It is implicit in the regnant social practices that she immediately recognizes the import of his behaviour. Of course it might have symbolic significance that he signals an interest in treating her gallantly – indications that she might possibly draw on in determining what his feelings for her are. But such an interpretive exercise is undertaken only after her already having understood the action uninterpretedly. It is different matter if we think of a man who happens to observe a young woman waiting at a bus stop. Her preoccupied and nervous pacing to and fro, anxiously watching out for the bus, leads him directly to interpret her behaviour: perhaps she is waiting for her boyfriend or is late for a dinner party. He forms a hypothesis – a hypothesis according to which her behaviour reflects a particular motive. We engage in interpretation when confronted with several representational possibilities and do not know with certainty which of them is correct.

It has been argued that acts are always description-dependent and have, in consequence, an indefinite or inconclusive character. This would mean, then, that the understanding of an action always involves interpretation. But that description-dependence which Elizabeth Anscombe has argued is scarcely any different from that referred to in the context of the theory-ladenness of other experiences. Actions are also perceived. And the interpreter needs to recognize the physical movement of the hand as a wave before she sees it as a wave. But if she is embedded in a particular social, linguistic or scientific practice she understands, *qua* participant in this practice, a whole host of things without any need to interpret them. There is little reason to believe that in this regard actions are any different from other experiences.

Interpretation in the natural sciences

Understanding in the natural sciences results not only from explanation but also from interpretation. Admittedly, the natural sciences rely heavily on causal explanations, while the humanities operate chiefly with structural, functional or interpretive explanations since the phenomena of interest to the human sciences are often representative of motives, desires, beliefs, etc. which are not facially transparent. But not infrequently we run into representational difficulties in science too, and so interpretation becomes a prerequisite for understanding, just as we engage in the identification of causes in the humanities in order to explain events in society and history. Interpretive explanations involve interpretations, but not all interpretations occur in the context of interpretive explanations.

Interpretation is endemic to natural science. At least four levels may be identified at which representation plays an important part in our description of the world, and at which interpretation would, if necessary, be able to forge connections to whatever is being represented. We shall refer to them as:

1. Natural effects and scientific data.
2. Formalism's connection to phenomena.
3. Models.
4. The metaphysical significance of theory.

Several of these have interpretive counterparts in literary studies. More on that later.

In many cases we infer what we cannot observe directly by focusing on what we can see. Sherlock Holmes is not alone in construing the presence of a bloody knife, two empty glasses, a full ashtray and a whiff of perfume on the air, as the traces of a particular criminal. Like Holmes we form hypotheses that construe indications occurring in our environment as the effects of particular causes. If, on returning to my office, I find my papers strewn across the floor, I surmise that a gust of wind through the open window has so dispersed them. (In America, I have been told, you never would have thought of a gust of wind but immediately interpreted the event as evidence that your office was burglarized.) Effects represent their causes and if we lack certain knowledge of the cause we frame a hypothesis – an interpretation connecting a phenomenon with a particular cause. The phenomenon is rendered intelligible. The cause explains the effect, but interpretation is needed to identify the most probable of the possible causes.

In every area of natural science the scientist construes phenomena as the result of particular causes. A trained particle physicist is able to identify straight off most of the tracks deposited on a photographic plate following exposures of the collision between an electron and a positron. Such familiar tracks are not normally the subject of interpretation. Their significance figures an integral part of the physicist's recognitional capacities. But if there are tracks that he has not previously met with, and which deviate from the norm, he will construe them as resulting from of a novel particle, or from of the aberrant behaviour of some familiar particle due to certain experimental difficulties. A good particle physicist needs continually to interpret the data yielded by his experiments, assessing these on the basis of anticipated results and conjecturing what the causes of any deviations might be. There are many sources of error in the highly complex experiments conducted in contemporary science where the available data are very far from being closely bound up with the items the experiment is designed to investigate. It is thus often not possible to determine offhand what the gathered data really represent. In consequence, scientific data have always to be interpreted in the light of various theoretical options.

Data and experimental outcomes are natural signs of their causes. There cannot, then, be an intended connection between these data and effects, and their causes. This is where the interpretation of physical data differs from the understanding of social and cultural data. In contrast to human acts, physical data are not meaningful in themselves because they do not stand in an intentional relation to their causes. But this does not prevent their being susceptible of interpretation. In consequence, (I) is not an exhaustive definition of all species of interpretation, but it needs to be modified to:

(I*) The connection between X and Y constitutes an interpretation for some person P, if and only if (i) P believes that X represents Y because X is in some manner attached to Y, and (ii) P's belief as expressed in (i) is presented as the result of a hypothesis.

How X is connected to Y is determined by the kind of objects being interpreted. If X and Y stand for physical phenomena it will be a case of cause and effect, but if they stand for items relating to human thought and agency, the connection will be intentional or conventional. Thus, there are two kinds of 'representing': causal, as when certain effects 'represent' the evidence for holding certain causes occurred, and non-causal, intentional, as in what a work of art 'represents'.

But it is not only physical data that is interpreted by the natural scientist. There are numerous intended relations in natural science which are also open to interpretation. A theory is a systematic representation of a segment of the world and the aim of science is to develop those representations that will enable us to explain our observations. Classical mechanics, electrodynamics, statistical mechanics, the theory of relativity, are all mathematical representations. Mathematics is a language we use to represent the non-mathematical. Whereas natural languages such as English, German and French are 'qualitative' languages that have application insofar as we view the world in terms of its qualitative properties, mathematics is one, or rather several, of the 'quantitative' languages at our disposal when we seek to configure the world in terms of numerical properties. Mathematically formulated theories may be seen as explicit determinations of linguistic rules for the unambiguous and exact description of things and events in nature. Newton's theory defines an exact language that makes it possible for us to speak of a wealth of mechanical phenomena in nature. In that role mathematical symbols refer in a highly specialized way to the phenomena we observe. The correlation is determined via a convention. There exists no given or natural way of setting it up. In Newton's theory for example, 'm' has been selected to designate an entity's mass, 'v' its velocity, and 'F' the force operative on a body.

Mathematically formulated theories do not differ essentially from natural languages in the way in which they represent the world. In both cases the coupling of linguistic expression and world occurs through the subscription to a convention on the part of the community using that language. The word 'book' might conceivably have come to refer to a table instead of a book, and the sentence 'The table is agaf' might have expressed the same meaning as 'the table is red'. Both natural languages and mathematics are used to represent the world in virtue of the rules governing the use of their respective symbols. So long as we use language and mathematics in accordance with the set rules and conventions there is no call for interpretation: as active language users we grasp their representational role. But once doubt arises as to the use of the rules in a particular case, interpretation is needed. *Qua* semiotic systems, language and mathematics refer to the world, and the setting up of a connection between signifier and signified only becomes the subject of interpretation if we are not already familiar with the rule or convention which instructs us as to how to do it.

Nor is it perhaps only the choice of linguistic terms that might be different. Could not the choice of the object the expression refers to also be different? Every representation rests on the abstraction, selection, accentuation and configuration of what is represented. We cannot describe all identifiable features at once, and there are many ways in which one thing appears to resemble another. All red things are alike, but they are also different from each other. How we describe a thing depends on what we want to convey by our description, but how can we be sure our description says what we want it to say? We must make a selection from among the contents of the world and isolate the features we identify as distinctive of those things we have set ourselves to describe. From among such features, we focus on just a few properties. Other properties we eliminate from our attention. Choice of description is also a choice of perspective. This applies particularly in the case of individual items. But it is by no means assured that there will always be just one correct way of describing the things we identify as being one specific type of thing. We shall return to this question in the next chapter.

In many cases, however, choice of description involves an interpretation of the phenomena to be described – the choice stands between alternative representations which are each and every one usable descriptions, and which, accordingly, do not yield a single unique or privileged account of the phenomena. We cannot always avoid interpretation in our choice of description by appealing to some convention or certain specified rules. The relation between designator and what is designated may be a connection between sign and a specific type of thing, but it could also be a connection between sign and an individual item that has to be interpreted if it is to be understood.

Newton's theory is precisely one such linguistic interpretation, whose applicability presupposes that we are able to isolate certain features of physical systems and describe them in a manner dictated by the rules. Newton applied a particular perspective to the world; he represented the solar system as a classic mechanical system and this enabled him to offer a good description in accord with the theory's terminological conventions. Among the manifold properties of the sun and the planets it was only their mass, velocity and the force of gravity obtaining between them that were necessary, so they were isolated and represented. Newton's perspective, his mode of representing a concrete physical system corresponds to the elaboration of a model. Einstein applied a slightly different perspective, representing the solar system as a relativistic mechanical system, describable in terms of his relativity theory. Newton's and Einstein's disparate representations of the solar system are at the same time two different interpretations of it. They each set up a hypothesis as to how the sun and the planets might best be represented.

The sun and the planets do not themselves instruct us as to how they are to be represented if we are to explain their movements and so come to understand such celestial phenomena. Neither of the two aforementioned representations presses itself upon us as the manifestly better candidate. Naturally, it may be justly objected that today we know that in certain respects Einstein offers a better representation of the solar system than Newton did. But as Bas van Fraassen points out in his discussion of the role of interpretation in natural science, in the eighteenth century Einstein's model would not have offered a representation superior to Newton's. At that time there were no phenomena that the one but not the other was able to describe. That came about only later. Bas van Fraassen's point is, then, that Newton's and Einstein's representations of the solar system are interpretations precisely because neither of them can lay claim to a unique or privileged status. If one were to doubt the validity of van Fraassen's point on the grounds that Newton's and Einstein's theories have not in the long run proved to be equally good options, such doubt would have to be deemed unjustified. For there are, today, alternative ways of representing the solar system which take a generalization of Newton's theory of gravity as their starting point. These alternative theories represent the same phenomena as Einstein's model. So even today Einstein's representation is just one of several possible interpretations of the solar system.

Different theories rely upon different interpretations. But one and the same theory can also give rise to alternative interpretations. Here I am thinking specifically of the theory's metaphysical meaning, the last of the four levels of interpretation listed above. Depending on how they are understood, theories and models all reach to a greater or lesser extent beyond what we are able to see with the naked eye. They represent in symbolic form a world which lies beyond the world open to sensory perception. Precisely because they represent things we are unable to perceive and attend to in a direct manner, what they can in fact be said to represent is not wholly determinate. Scientific theories and models are distinguished by a range of abstract structures and properties which the scientist cannot grasp without framing an interpretation that specifies what, if anything, these structures and properties indicate about unobservable items.

Natural science may be regarded as an exact ontological account of what exists in the world and the principles governing it. What it tells us about it is based on observation and experiment. But since such 'worldly' phenomena are brought about by things we cannot observe, and our theories and models represent these things in very abstract and theoretical languages, we need to understand what these descriptions really mean. We need to have descriptions put in the context of other descriptions that we already understand; we need to have them cohere with our account of everything else in the world.

Quantum mechanics – the theory of atoms and the smallest constituents of nature – demonstrates more clearly than any other theory how in natural science understanding involves interpretation. The world of atoms strikes us as very alien because it has properties that larger things do not, and because we can speak of them only in a mathematical language that represents observational properties in an abstract vector space. Quantum mechanics explains the observable by using a language which purports to refer to the unobservable. Today there is a wealth of interpretations of quantum mechanics all of which, in various ways, advance metaphysical interpretations of the world of atoms that extend far beyond what can be observed.

Among the most prevalent interpretations is the 'Copenhagen interpretation' which derives in large part from Niels Bohr. This interpretation argues against the view that quantum mechanics represents a world over and above that which can be observed. But most other interpretations conceive of quantum mechanics as a theory offering a quite distinctive interpretation of reality. One such is the many-world interpretation that sees the values of the quantum mechanical probabilities as realized in each of their respective worlds, quite separate from each other. Another is the 'decoherence interpretation' in which the theory can be said to represent the world in terms of alternative histories each with a probability determined by the distribution function all superposed as the ontologically fundamental entities. Yet others again include the 'fuzzy interpretation', the 'modal interpretation' and 'Bohm's quantum potential interpretation'. What all these interpretations have in common is that they advance a hypothesis concerning what the mathematical formalism represents, and by doing so they seek to get a grip on the theory's ontological implications.

Several of the interpretive levels we encounter in natural science we meet again in the humanities. In literary studies we make a distinction between what we might call textual and literary interpretation. Textual interpretation is directed at the understanding of the text in terms of its literal, linguistic meaning, while literary interpretation seeks to establish an understanding of the work's meaning *qua* literary artefact. To the extent that, as language users, we are able understand a given text simply by reference to the rules and conventions of the language, other meanings get no foothold, and so we have no problem in making sense of what we read. But if we do encounter difficulty we have recourse to interpretation. The counterpart in natural science lies, *mutatis mutandis*, in the indeterminacy of the relation of a theory to the world, or the fact that various models may be framed, each of which purport to represent one and the same phenomenon. The literal meaning of a text may be vague and ambiguous, and the author's semantic intentions as regards form and content may be described in a number of ways. Diverse literary interpretations of texts are thus comparable with differing metaphysical interpretations of the materials of natural science. Such interpretations offer readings of texts that, besides the immediately given meaning, carry symbolical or metaphorical meanings, reaching beyond the fictive world that emerges through the literal reading.

Here the question might be raised as to whether it is not precisely the task of natural science to bring about unity and clarity in our understanding of nature by eliminating indeterminacy and ambiguity, whereas the human sciences are expected to attempt to show that beliefs, texts, paintings and acts may have multiple meanings and are open to many interpretations. Indeed, it might be said that the human sciences bring out the value of art as residing not least in its openness to an array of modes of understanding. A good work of art opens to diverse interpretations, and so it would not be an appropriate charge to lay at the door of the humanities that they do not proffer finality. This question has, however, several strands to it that are worth keeping separate.

To begin with, there is the fact that in every representation, irrespective of whether it is of nature or of meaning, there will be elements which are not unambiguous or fixed by convention. Any representation is always open to reinterpretation with respect to what it represents if some relevant phenomena turn out not to be properly represented by the representation.

Second, it has been shown both logically and historically that it is possible to set up a plurality of representations, which is to say various descriptions of the world, that are empirically underdetermined. This entails that observational data cannot be used to decide between one theory and another. So the popular conception of natural science as a collective enterprise that delivers uniquely right answers to our questions about the natural world is an ideal which cannot be fulfilled, even in theory.

Third, there is much to suggest that nature does not even always return unambiguous answers when we put to it unambiguous questions. In the atomic world we encounter entities displaying the behaviour of both particles and waves, and we are constrained to deploy two different representations to give an exhaustive description of such experimental phenomena. We have to set up two models of atomic processes which are mutually exclusive but which each display features of the phenomena we observe. Such were the incompatible representations that Niels Bohr called complementary descriptions.

Fourth, and finally, it might be said that alternative representations do not have to be considered a negative but may also be seen as advantageous to science. In many situations ambiguities and a certain measure of gappiness will lead to new discoveries and a deeper insight into nature. Thus Paul Feyerabend has sought to construct a methodology where working with alternative theories figures as an all-important principle. Not only are there historical examples of the discovery of novel phenomena owing to the availability of alternative theories, but also there are phenomena that would never have been discovered had a plurality of representations not directed our enquiries. If natural science were not, in the same way as human science, open to new interpretations of phenomena as much as of theories there would scarcely have been any scientific progress worth talking about.

No doubt there will be readers who are ready to concede that interpretation is a feature of both natural science and of human science, but who will still feel that there is a vast difference between the interpretations in each case. By way of response I would say: both Yes and No. I answer No, if it is claimed that there are different kinds of interpretation. The structure in all interpretations is the same. There are no constraints on interpretations that go beyond those articulated in (I*). Interpretation is not confined to meaning. However, my answer is Yes if instead of different species of interpretation, what is meant is that the interpretations are of generically disparate objects. If what is represented is meaningful, then of course it must be interpreted as such; and it is only understood when understood *qua* meaningful. Understanding in the human sciences is different from what it is in natural science, because the subject matter in each case is not the same. It is always what is to be understood that is different. But this does not make understanding in the humanities more intuitively intelligible, deep or right. How (or whether) I empathize or find intuitively intelligible, say, someone's behaviour is dependent on how I interpret it, but the interpretation depends in no essential way on my empathy or intuitions. Therefore I conclude empathy and sympathetic identification must be removed from the philosophical analysis of understanding and interpretation.

References

Faye, J. (1991), *Niels Bohr, His Heritage and Legacy*, Kluwer Academic Publishers, Dordrecht.
Fayerabend, P. (1975), *Against Method*, NEP, London.
Taylor, C. (1981) 'Interpretation and the Sciences of Man', in *Review of Metaphysics 75*, pp. 3-51.
van Fraassen, B. & Sigman, J. (1993), 'Interpretation in Science and in the Arts', in G. Levine (ed.), *Realism and Representation*, University of Wisconsin Press, pp.73-99.

5 Facts

A striking difference between natural science and the humanities leaps to the eye immediately. Natural science deals with objective facts whereas the humanities are typically not confronted with a mind-independent reality. In consequence, it is commonly thought that natural science and the humanities each have their own distinct methods because the ways in which they produce explanations are sharply different. I consider this to be a mistaken conclusion. The fact that natural science looks for causes and that the humanities study meaning and understanding is not a difference grounded in any contrast between objectivity and subjectivity. This remark must not be misunderstood. Meaning and understanding are indeed mind-dependent, but that does not preclude their being based on objective facts.

Facts are what we explain and interpret. In science every explanation is given by reference to a hypothesis, and every hypothesis will be an interpretation of facts. Let us begin by looking at facts as they figure in natural science. A hypothesis is framed on the basis of a collection of observable individual facts, and is subsequently used to explain other facts of the same sort. When we want to explain why iron rusts after exposure to wind and weather, it is the *fact* that iron rusts that we want to explain. But are there really *objective* facts in natural science?

Today it is almost philosophical dogma that all observation is theory-laden. The strong version has it that our theories determine what we observe and, in consequence, what the world is like. Such a radical formulation makes it impossible to speak of iron rusting outside a theoretical framework. According to that line of thought, not even natural science boasts objective facts – not in the sense of facts obtaining independent of the way we describe them. Both in natural science and in the human sciences facts seem to be determined by our theories and our epistemological interests. The theory-ladenness of facts in natural science naturally does much to cancel out the difference between its purported objectivity and the alleged subjectivity of the human sciences. But even if facts in the natural and the human sciences are thus shown to be on a par, as we shall see, there is a crucial difference all the same. Once we have taken a closer look at what we understand by 'facts' in natural science, we shall be in a position to address the issue of facts in the human sciences. So we will return to that topic at the end of the chapter.

The descriptive position

There can of course be no doubt that the way in which we describe entities and their properties reflects a conceptual and linguistic framework of description. To that extent our observations are indeed theory-laden, inasmuch as we cannot make the observation that iron rusts without possessing the concepts 'iron' or 'rust'. Nevertheless, the idea that it should not be true that iron rusts irrespective of whether we are in possession of a concept applicable to that process strains credulity. The process goes on quite independently of our ability to register it.

We have already several times referred to the fact that whether we just see something, register a person's behaviour or read a text, we grasp what we see, register or read as something constituent in immediate awareness. We do not need to infer it from something more fundamental. If I observe a car, say, then I just see it straight off as a car; I do not need to start interpreting my sense impressions before being able to see it as that kind of motor vehicle. This is because my concept of a car, my linguistic competence, is a constituent in my visual perception of the car. But it also seems correct to say that the reason why I have a concept of a car is that cars exist, and that I am able to recognize them such in virtue of possessing the appropriate concept. My observation consists, then, in the acquisition of the belief that what I have presently before my eyes, a car, falls under the concept 'car'. We can formulate this as follows: the observation of some object O as a specific exemplar N, consists in the acquisition of the belief that the conditions for the application of the word 'N' to O are satisfied.

Now we do not just see things without seeing them as possessing specific properties. I see that the car is red because it is a fact that it is, and a fact that induces in me the belief that the conditions for the correct utterance of the sentence, 'The car is red', are presently fulfilled. Facts are expressible in declarative sentences and are what render such sentences true or false. I observe a fact if, via my senses, I am invested with the belief that some sentence, 'N is P', denotes what I observe. But what is it that leads us to say of things and facts that they exist irrespective whether we have concepts applicable to them, but which also makes it possible that we are able to use language correctly?

One attempt to solve the problem is of course to say that the subject matter of natural science is *natural kinds*. Nature is the way it is even though we might believe it to be different from the way it is. Nature's inventory already belonged to a range of different categories before humans existed and began to understand it; each thing possesses certain specific properties regardless of how we set about investigating them. Besides the four forces by which it is governed, nature encompasses atoms, molecules, iron, copper, methane, stars, planets, beeches, apple trees, toads, storks, lions, tigers, and of course human beings. All entities, and all matter in nature, appear to be divided into particular kinds into which all like things, or all like matter falls. In the natural kind category 'lion', we find all lions, in the natural kind category 'gold', we find all instances of gold and in that of 'atom', we find all atoms. So if these natural kinds exist there must perforce be objective facts. An item that belongs to a certain natural kind will have particular properties, and it is an objective fact that it does.

But how do we define a natural kind or how do we identify a natural division? According to the descriptive theory we can isolate three aspects of the concept 'natural kind'. The term denoting any natural kind has:

1. *Meaning or intension.* A definition or a definite description.
2. *Extension.* All the individuals that have the properties determined by the intension.
3. *Reference.* An abstract entity or a universal.

Each of these aspects contributes to the determination of what a natural kind is and, in virtue of that, what objective facts are true of it.

Meaning is the description of those properties a thing must have to be a thing of the kind that it is. Take a tiger, for example. To be a tiger an animal must be a mammal, a predator, tailed, four-legged, tawny yellow-coated, marked with black and white stripes etc. These various properties define what it is to be a tiger, and their description fixes the meaning of the term. It is this meaning that an individual knows when, *qua* competent language user, he is able to use the term 'tiger' correctly. We call this meaning 'intension'. Now the 'extension' is the set of all animals that has those properties, and only those properties, that are determined by the intension. It comprises all the tigers in the world – those presently existent as well those once existent and those yet to be born. Finally, the reference of the word 'tiger' points to the natural kind, tiger. The referent is the abstract entity, 'tigerhood'.

It has long been recognized that there are considerable problems attaching to this descriptive theory. For an albino tiger is also a tiger – a tailless albino tiger no less – and a tailless albino dwarf tiger, still no exception. But such specimens fall outside of the definition of 'tiger', which has it that for something to be a tiger it has to be a large, long-tailed, black-and-white striped, tawny yellow animal. Further, consider dogs. Scarcely any other animal varies so much in size and appearance as does man's best friend, spanning the large black-dappled Great Dane and the Miniature Schnauzer. The differences here are far greater than those separating closely related species. It will rightly be asked what the properties are that are the defining characteristics of dogs. The theory as originally conceived seems wanting.

Little wonder is it, then, that it has undergone modification at the hands of philosophers. One alternative is the cluster theory, holding that none of the properties associated with tigers are necessary. It is enough for tigerhood that the beast in question has just some of the properties figuring in the definite description. The extension contains the set of animals that have most of the properties. No unique property of those named is necessary, but suitably many are sufficient to make something a tiger. But what is 'suitably many'? The theory does not say.

The cluster theory also can be shown to be untenable. Think of an android – the exemplar in *Star Trek*, say. In the future, researchers may be able to create a humanoid robot that displays all the external features whose description is assumed to be sufficient to pick out a human being. It will be able to talk, give answers and put questions. But we would not say that such an android is a human being, for it has neither heart nor liver – and nothing remotely resembling an emotional life.

However, the descriptive theory faces even worse problems. The properties of natural things come to light only gradually. Science is continually increasing our understanding of the properties of natural things: further items of knowledge accrue to what we already know, items not represented in the original definitions. Today we know that water consists of the molecule H_2O. But water was originally defined as a colourless, odourless, fluid in which sugar (but not wood or iron) dissolved; water is a liquid that can evaporate and freeze to ice. So an empirical discovery has enabled us to give a new definition of water. The old and the new concepts have the same extensions, but different intensions. In the same way, gold was once defined as a shiny, heavy, yellowish, malleable, metallic substance with a melting point of 1063° C. Today we know that gold has the atomic number 79, *i.e.* gold has 79 protons in its nucleus. A new definition of gold draws on new discoveries to establish another meaning. So, all definitions seem to be open to continued expansion or revision. A name for a natural kind can change its meaning or intension so long as its extension remains the same. But had the descriptive theory been correct, the retention of the original meaning of the name would have been essential to its proper use.

The essentialist position

A distinction between *nominal essence* and *real essence*, stemming from Locke, may serve to shed light on some of these difficulties. The nominal essence comprises the properties a thing must have to be called a particular kind of thing. The nominal essence is, then, what is described by the term's intension. When an individual is able to use the name of a particular type of thing correctly, it is because he recognizes the thing's nominal essence as that which members of the linguistic community associate with the use of the word. The nominal essence of gold includes the fact that it is a lustrous, heavy, yellowish, malleable, metallic material with a melting point of 1063°C.

Natural things also have a real essence, as each and every thing must have – not in order to be *called* a particular kind of thing – but to *be* a particular kind of thing. The real essence makes the thing what it is. The real essence of gold is the existence of 79 protons in the atomic nucleus. The difference between the two types of essence is a difference between what is needed if we are to be able to recognize a thing as such, and what, from the hand of nature, must hold for the thing to be what it is.

Nominal essence is epistemically necessary: An object G has the property E with epistemic necessity if it is not possible to envisage a situation in which G does not have property E without our losing our ability to identify or recognize G.

Real essence is ontically necessary: An object G has the property E with ontic necessity if it is not possible to eliminate E from G without G ceasing to be the object G.

Interestingly enough, it is here that there is a parting of the ways for the descriptive and the essentialist theories. Natural kinds have a nominal essence which, according to the former theory, determines the definition of the name of the species, and which is, in consequence, semantically necessary. The loss of the nominal property E prevents us from using the name 'G' in accordance with its definition. The latter position disavows any such necessity.

Natural things have no nominal essence, only real essence. This claim is at the core of the essentialist position. Nominal essence is neither necessary nor sufficient for the notion of a natural kind. Other things, such as, for example, articles of everyday use, may indeed have a nominal essence but no real essence. A hammer has a nominal essence: it is a tool with a heavy blunt head of metal designed to drive in nails. Such descriptions are ones we might call nominal. It is not the real essence of a hammer to drive in nails. A hammer does not cease to be a hammer if it cannot drive in nails. It is merely broken.

The contemporary version of essentialist theory derives in the main from two American philosophers, Saul Kripke and Hilary Putnam. Each of them has argued that the descriptive theory is untenable. Kripke has shown that a nominal essence is not necessary to fix the name of a natural kind. The determination of extension does not require reference to intension. An example will show why. An explorer returns home after a trip to the Himalayas where he has observed a nine-foot tall, hirsute, bipedal, anthropoid figure, which the locals call 'yeti' and which he calls the abominable snowman. It apparently lives hidden away among boulders in the perpetual snows of the mountains. Later another explorer emerges and relates that he has seen traces of the abominable snowman in the snow. A third sets out and gathers accounts of its appearance and habits of life that match the reports of the first, while yet a fourth finds evidence that the abominable snowman lives on yaks. All are agreed that the abominable snowman is an anthropoid ape that survived human evolution through living in total isolation from the wider world.

Now when a group of explorers next makes the trek to the Himalayas in search of living specimens, it turns out that high up in the mountains there lives a hitherto unidentified species of bears that occasionally raise themselves up on their hind legs and walk about, and it is not difficult to imagine that it is these beasts that the locals call 'yeti'. The nominal essence of this species cannot be that it is a nine foot tall, hirsute, bipedal anthropoid creature. There is no nominal essence that both the 'yeti' has and that the locals believe it to have. In consequence, its nominal essence cannot be definitional of the name 'yeti'. For all that, the locals are talking about the same creature after the discovery of the new bears as they were before. The name is used exclusively of the yeti. So it cannot be the putative intension of the name that determines its extension. The example shows that nominal essence is not part of what determines a natural kind. We can assign a name to a natural kind without that name having a determinate meaning.

For his part, Putnam argues that the nominal essence is not sufficient to fix the use of natural kind terms. Intension is not sufficient to determine extension. Consider, says Putnam, two worlds, the earth and a twin planet, that are as indistinguishable as any two planets could be. What we call water here on earth is ostensibly also called water on Twin Earth. In both places water comes out of the taps, quenches the thirst of animals and humans, is used to water plants, freezes to ice at 0°C. and boils at 100°C. There is simply no perceptible difference between water on the one and the other planet and thus no difference in the beliefs of the inhabitants of Earth or those of Twin Earth as to what water is.

But there comes a day when chemists on Earth discover that water consists of H_2O, while their counterparts on Twin Earth discover that water is really XYZ. There is a difference, then, between water on Earth and on Twin Earth. The word 'water' does not have the same extension on Earth as it does on Twin Earth. The supposed nominal essence is no guarantee that we, and our Twin Earth doppelgangers, are talking about the same thing, and so it is not crucial to our use of the names denoting natural kinds.

If nominal essence does not determine a natural kind, then what does? Nothing other than real essence, but not in the form of a new meaning which defines what it is to belong to the kind in question. Beyond this point Kripke and Putnam part company.

Kripke argues that extension is fixed via its causal reference and that the names of species and substances are rigid designators. This theory is known as the causal theory of reference. The name of a natural kind is associated with the item to which that name is applied through an act of baptism, analogous to those instances in which a boy is given the name, say, of 'Simon', and a girl the name, 'Matilda'. The act of baptism itself initiates a causal process: every subsequent language user's use of the proper name 'Simon' or 'Matilda' of a particular individual is determined by the parents' ceremonial conferral of that name on their offspring, and not of the names' intensions. For they have no intension.

By the same token, argues Kripke, the linguistic community uses the name 'tiger' in accordance with the first language users' classification of tigers. Language has a history. It is of the nature of language acquisition that its history is perpetuated via the causal influence of the proficient language user on his less proficient counterpart: the latter is led by the former to apply the term in conformity with the original baptism or dubbing ceremony. Any such name is a rigid designator because it refers to its objects in a way that makes no reference to their particular properties. The constant in the objects referred to is the real essence.

Putnam tells a different story. There exist 'archetypes', concrete examples that we start off by associating with a particular term. This use is then extended to other exemplars. The extension of the name is fixed through a 'same-type' relation in which only things of the same type participate. The idea is that whatever matter shares the same general physical properties such as freezing point, melting point, density, proves to be composed of the same chemical constituents, and so enters into a 'same-type' relation which identifies all such samples as samples of one and the same substance. And that relation is one shared by them even if we have no knowledge of the fact. Think of a raindrop – our 'archetype' of water. The raindrop has determinate physical properties such as freezing point, boiling point and density, all of which can be explained by its possession of the underlying microstructure, H_2O. Therefore, water now becomes, in consequence, everything that consists of molecules with a certain number of hydrogen and oxygen atoms. Same-type relations subsist between real essences. The word 'water' will never come to denote anything other than that which stands in this relation to the original raindrop. It will never pick out XYZ. The task of science, says Putnam, is to discover these same-type relations.

However, the difficulties attaching to Putnam's proposal are quite considerable. For what is it to be 'of the same kind'? As has already been noted, things resemble one another in many respects, and differ from one another in many respects. Is it not the case that we refer to natural kinds precisely in order to indicate wherein the relevant similarity consists between individuals and elements that fall into the same class? For example, tin occurs in 21 different isotopes, *i.e.* types of tin with the same number of protons but with a different number of neutrons in the nucleus. How do we establish that it is the number of protons and not the number of neutrons that pertinently marks tin out as a natural kind? It seems to be possible only in virtue of our prior possession of a concept of tin that corresponds to its being the number of protons and not the number of neutrons that determines the extension of this concept. Had that not been so, we could equally well have used the number of neutrons to enumerate 21 natural kinds rather than opting for one.

By the same token, one might point out that even though chimpanzees and human beings are very different, 99% of their genes are the same. If DNA sequences are what determine the species to which an individual belongs, why is it the slight genetic difference between chimps and humans that is crucial to species-membership and not the vast genetic inheritance that they share? Moreover, the genetic endowments of both chimps or of humans are not in each instance exactly the same. Some people are blue-eyed, others brown-eyed. Some people have dark hair, some are red-headed while others again are fair-haired. Here the relation is the reverse: it is the genetic inheritance that we share rather than the differences that separate us that determines the natural kind to which the individual belongs. We may observe again, then, that what counts as a relevant comparison is wholly determined by the fact that we already possess an understanding of what we are talking about.

The criterial theory

The essentialist theory is of interest because it explains our use of natural kind terms in areas where the descriptive theory falls short. Unfortunately the essentialist theory evinces weaknesses at points where the descriptive theory appears to be strong. In some cases we are forced to refer to something resembling a nominal essence to fix the extension of a name – in other cases not.

That the atomic number of gold is 79 is an empirical discovery. In establishing that fact physicists must have already had a concept of gold that they could unequivocally rely on in establishing that precisely gold had that atomic number. And what could that concept have been other than the nominal essence of gold? Had physicists not been relying on gold's nominal essence in determining the extension of gold, how could they possibly have discovered that *all* gold has the atomic number 79? The atomic number is only of interest because it is shared by all gold. Moreover, taking into account that the atomic number is an empirical discovery, it is possible that physicists will one day discover that gold does not, in fact, have the atomic number 79, but 78 – or else, perhaps, the idea of atomic numbers might conceivably be abandoned altogether. On this scenario the gold to which we refer could not be the substance with the atomic number 79, for it would not exist. The only thing physicists would have to fall back on is something analogous to the nominal essence of gold which furnished them with the concept in the first place.

Let us return to Putnam's Earth and Twin Earth, but now let the likeness between them be not the 'nominal essence' but the real essence. On Earth H_2O is water, as we know it, but on Twin Earth H_2O is a hard dark brown substance of which mountains are composed. This means that water consists of H_2O in both worlds but each has its own 'nominal essence'. This is surely a point against Putnam. On this scenario 'water' cannot have the same extension in each case. For now there is no connection between possession of the same microstructure and possession of specific properties that can function as criteria for the correct use of the word. Reference, however, is not objectively embedded in the nature of things: it is contingent upon it being epistemically possible for us to demonstrate the existence of a referent. The criteria for correct use need to be observable if they are to determine reference. 'Water' does not mean H_2O. In consequence, there must be something over and above the fact that 'water' refers to H_2O that picks out what that name correctly refers to, and it must be such that when we use it as a criterion for correct use, the name does in fact refer to H_2O. But that is precisely not the case if water has the same real essence but quite different empirical properties in the two worlds.

It is human subjects who establish the reference between names and objective reality. First, the linguistic community determines the conditions under which a given name or predicate properly has a reference: in doing so the community determines the criteria that must be satisfied for something to be designated as the referent of the term in question. Second, it is only through observation and the ability to manipulate the environment that we are able to determine whether the criteria for reference are in fact satisfied. As a corollary, it is also the case that the criteria for the reference of a given term must, in some sense, be epistemically accessible; in general terms this means that the satisfaction or otherwise of the relevant criteria must be observable or have some observable effect. Until we have ascertained whether the conditions for possible reference are realized, a name cannot meaningfully be said to refer to an objective reality.

Nonetheless, Kripke's case for the causal theory of reference remains persuasive. His arguments might well be found infelicitous by many on the grounds that chains of communication are not the same as causal chains. A parrot that has learned to say 'sugar' does indeed figure in a causal chain going back to the original dubbing, but it does not, on that account, grasp what the word refers to. The causal connection, then, is not sufficient to establish reference. But there can scarcely be any doubt that it is a necessary condition – and not least as seen from the criterial perspective.

We are only able to observe and manipulate the physical world because we stand in a causal relation to it. A correct theory for names and designators of natural kinds would seem to have to satisfy two conditions:

(i) There is a causal connection between the use of the name and its bearer.
(ii) The causal connection is determined by the criteria we have elected to use to identify the bearer of the name. The criteria are the bearer's sortal properties.

For example: Iron is a natural kind and it is one of the sortal properties of iron that it rusts. We shall call a property 'sortal' if it assists us in the identification of a natural kind, and we shall call the predicate 'rusts' a sortal predicate because it enters into the criteria that an entity must satisfy if it is to be a bearer of the name 'iron'. A melting point of 1530°C., magnetism, silver-whiteness, disposition to rust etc. are accordingly sortal properties of iron.

Now it might strike the reflective reader that our analysis of the concept natural kind is now approaching the point reached in our analysis of motives and actions. If we apply the criterial analysis to natural kinds, we shall be able to describe the relation between the name, the criteria for the correct use of the name, and the items to which the name refers, under three points:

(a) The connection between criteria (*i.e.* the sortal properties) and what they are criteria for is causal, which is to say the criteria are, logically speaking, neither necessary nor sufficient. The connection is contingent.
(b) The fact that there is a causal connection between criteria, and what they are criteria for, is a logical fact. This means that it is part of the meaning of a natural kind, which is introduced or explained, that the sortal properties are good evidence of the name of this kind being correctly applied.
(c) The criteria are defeasible.

It is, then, part of the concept 'tiger', a constituent of the term's meaning, that certain causal properties of a suitable kind (sortal properties) constitute good evidence for the existence of tigers. The causal connection between tiger and its sortal properties enters into the definition of 'tiger'. In other words, we can say that reference to the sortal properties of things constitutes the criteria for the correct use of the name of the item in question, and that these properties thus determine what the items are to which the word refers. The relation between the relevant sortal properties and the criteria for the application of the name is the same as that between an action and a motive. Often the action itself will function as a sortal property of a particular motive when we use it to identify the attitude, desire or intention of the individual in question.

It is important to make note of one specific aspect of the above account of natural kinds. We have been saying that sortal properties constitute *good evidence* for the identification of the bearer of the name. The problem is, however, that good evidence offers no *guarantee* that we are referring to the same thing. In principle we might be mistaken. In contrast to what the descriptive theory asserts, according to the criterial theory, sortal properties (the nominal essence) are neither logically necessary nor logically sufficient to determine the extension of the name. Sortal properties do not function as essential, but as causal properties.

Putnam's Twin Earth argument showed us that the presence of sortal properties is no guarantee that we are referring to the object which we believe we are referring to. But that argument's point also rests on the denial of any causal connection between the possession of a particular microstructure and a particular way things look. If H_2O and XYZ do not look different from each other, what makes it possible for us to ascertain that the one is H_2O and the other XYZ? If there is an ascertainable difference in the real essence, there must also be a difference in the sortal properties of things. For if two real essences have the same effects we have no means of telling them apart, and consequently have no reason for saying that the one consists of H_2O and the other of XYZ. So sortal properties must be *epistemically* necessary and often *epistemically* sufficient to be identified as H_2O and as XYZ (for example as alcohol – CH_3CH_2OH).

When Kripke states that nominal essence is not logically necessary he is of course right. But the reach of the example we have looked at is limited. A number of the properties that are in the first instance attendant on the bearer of the name 'yeti' are not properties that define the meaning of the word and so determine its extension. But this is due to the fact that none of these properties is causally connected to what it is to be a yeti. But there are other properties displayed by the yeti both prior to, and subsequent to, their being identified as bears, which contribute to the determination of causal reference. These include its habitat and habits of life, its height when standing on its hind legs, its brownish coat, mammalian status etc. These features give us an assurance that the zoologists are referring to the same animals as the local population and the first explorers.

We must have visual markers if we are to be able to identify the things we talk about. The fact that sortal properties function as criteria for the reference of a name means that it is possible for us occasionally to be mistaken in referring to them, but not systematically and invariably. Sortal properties are in fact all we have to go by in identifying the bearer of the name, and so must determine its correct application. We cannot, that is, revise our belief that named properties fix the term's reference, while simultaneously revising our belief regarding what it is the word stands for. If we were one day to discover that iron does not have the atomic number 55, we should perforce still be in a position to identify iron – which now does not have the atomic number 55 – by reference to its sortal properties. In that situation sortal properties function as epistemically necessary criteria: nothing that fails to display certain specific sortal properties will be identifiable as iron. And should we at some point come across a metal which, like iron, has the atomic number 55 but otherwise has other quite different properties, we would not conclude that the sortal properties of iron are not a reliable indicator of iron but, rather, that atomic numbers have no causal implications for the appearance of metals. The conclusion would be that atomic numbers are scientifically without their current interest because they do not indicate the type of metal to which a sample belongs.

There obtains, then, a definitional relationship between the name of a particular type of entity, and the criteria that constitute good evidence for the existence of that entity. We understand the name 'tiger' in that we understand that the sortal properties of the tiger give us good reason to identify something as a tiger. That circumstance constitutes part of the meaning of 'X is a tiger' and is therefore analytic and knowable *a priori*. The other component of the meaning consists in the fact that the concept 'tiger' refers to an animal with the specific genes that make it a tiger.

(D) X is Y, if and only if X fulfils the criteria for X being Y (*i.e.* X satisfies the sortal properties a, b, c, d, e which are good evidence for Y) and X thus refers to Y's 'real essence'.

But it is important to emphasize that in the first instance the relation cannot be knowable *a priori*. The relation is analytical *a posteriori*, rather. Which sortal properties attach to a natural kind is discovered through experience. It is only gradually that we come to recognize which properties constitute good evidence of something's being iron. For instance, it was only in the nineteenth century that it was recognized that iron has a melting point of 1530°C., just as it was not until the twentieth century that the 'real essence' of iron was identified. Moreover these properties of iron are an objective fact. After we have established an analytic relation between a name and certain predicates, then we can gradually admit new sortal predicates into our determination of the name's application, if we come to recognize that the referent of the name has sortal properties of which we were previously unaware.

Analytic statements normally express either tautologies or synonymities. How should their truth, then, be knowable *a posteriori*? Since Kant, statements have been divided into those that are *analytic* and those that are *synthetic*. Analytic statements are those that are true in virtue of their meaning and which are thus cognized *a priori*. They do not tell us anything about the world, only about how language is used. This applies to a statement like 'All bachelors are unmarried men': we do not need the benefit of experience to discover that it is true. It is true by definition. Synthetic statements like 'The tree has no leaves on it' are about the world, and such sentences are, in consequence, only true if there is something in the world that makes them so. Their truth can only be apprehended *a posteriori*.

The truth of a definition such as 'A bachelor is an unmarried man' can thus be ascertained *a priori* because the identifying criteria, that of being both a man and unmarried, are *infallible*. Such infallible criteria pick out bachelors in all possible worlds. Conversely, the truth of a definition such as 'Iron is an element with a metallic lustre, malleability, a melting point of 1530°C., …, and an atomic structure containing 55 protons' would not be knowable *a priori* because the criteria included in the definition are *fallible*. They do not pick out iron in every possible world and so the name is not a rigid designator. The currently accepted definition has been empirically 'discovered' bit by bit, and the criteria, the sortal properties used to identify iron may, in principle, be again eliminated from the definition if experience warrants so doing.

Sortal predicates are linked to a given name through the initial association of that name and a host of predicates with one and the same extension.

78 *Rethinking Science*

The term 'iron' originally attached to a lump of matter having certain characteristic properties that lead us to dub the matter 'iron'. It is also through experience that we establish that every lump of matter denoted by the term shares not only many of the same type of properties, but also other types of properties than those that occasioned our so dubbing it. We come to recognize, so to speak, which possible predicates collectively belong to a criterial description of the matter in question, and which do not. Only then is there an empirical identification of the bearer of the name with a criterial description. The name and the criterial description that link the bearer of the name to certain properties have, henceforth, the status of extensional equivalence (the same extension in every possible world) but not necessarily intensional equivalence (the same intension in every possible world). Extensional equivalence comes about through the properties in question initially being regarded as sortal, and so as epistemically necessary for the identification of any bearer of the name. This is followed by the semantic identification of the name with the criterial description and with what that description is a criterion for. The name thus becomes synonymous with the expression specifying the set of sortal properties and the 'real essence'. Subsequently the name figures as intensionally equivalent to the criterial description and the 'real essence'. It is, then, ontologically necessary for an entity that it have precisely all its nominal properties, observable as well as unobservable, in order to be an entity of a particular kind or a particular kind of matter.

Eventually it became part of the meaning of 'iron' that it has atomic number 55, and similarly of 'water' that it is H_2O. The 'real essence' of iron and water began to function as (part of) the 'nominal essence' for the linguistic community of physicists and chemists.

Facts in the human sciences

There are, then, in natural science, objective facts that cannot be said to be theory-dependent in any problematic sense. They comprise those expressed in analytically *a posteriori* sentences about the relations between things and their sortal properties. Things have the sortal properties they do, but our beliefs as to what is a sortal property may of course be mistaken. For there are a host of other facts about iron, say, which do not concern its sortal properties, but concern its *accidental* properties, and which are expressible in synthetic *a posteriori* judgements.

But, more generally, I would contend that nothing figures as a truth-making fact until it has been identified as such. To that extent *all* facts are theory- or mind-dependent. Facts are what make sentences true – facts are truth-makers and sentences are truth-bearers. A sentence such as 'Penguins live in Antarctica' is true if and only if it is a fact that penguins live in Antarctica.

We use language to represent reality. Such representation is conventional – it does not mimic reality. The sentence, 'Penguins live in Antarctica' is true not because it pictures reality but because we have adopted conventions determining what counts as the appropriate sort of fact, and because a given concrete instance of it makes that sentence true. We can put this in another way: a certain sentence type identifies a certain fact type, while individual tokens of a fact type are what make token sentences of the corresponding sentence type true. The extent to which a concrete fact makes an individual sentence true or false is thus an objective, mind-independent circumstance – it is not dependent on me or on any other individual whether the sentence, 'Penguins live in Antarctica' is true or false. What is partly theory- or mind-independent is whether, in the first instance, there is a fact type that corresponds with a certain sentence type. We need to have determined the empirical conditions that must be satisfied for such facts to obtain before we can say that individual statements about particular facts are true or false.

In order to be able to establish a convention governing the link between language and reality it must of course be possible to observe a language-independent fact without having to see it as a particular fact. There must be a causal connection between the use of words and that to which we apply them set up by those dubbing ceremonies that establish correct usage. But the causal connection between language and reality comes about via an *intentional* act. The causal connection between the use of words and what they stand for does not rest exclusively on a natural or physicalistic/physical process. It comes about through the setting up of a correlation between a certain sentence type and a certain fact type – a state of affairs – observable by us. A fact figures as such only when so designated by a knowing subject.

That said, the question naturally arises as to where this places natural science in relation to the humanities. I would respond to that by saying that there exist analogous, partially objective, partially theory-dependent facts within the human sciences, and it is these facts that we seek to understand when conducting research within the human and social sciences.

To avoid misunderstanding let me add that I am not contending that there are natural kinds in the field of objects addressed by the humanities, perhaps with the exception of psychology. The humanities work exclusively with *nominal* objects: artefacts or products that human beings have created. In contrast to natural objects, cultural entities would not exist, had human subjects not first devised them. No artefact exists entirely as a physical object. Some of them exist primarily as physical objects, such as a hammer, whereas others exist primarily as non-physical, intensional, or mental entities, such as a poem.

Thus, natural objects differ from cultural entities inasmuch as the criteria for a natural kind offer only *good evidence* for the correct use of the name while criteria for a nominal class *warrant* the correct use of the name. Or, put differently: sortal properties attaching to natural kinds are causal properties, but in the case of nominal entities they are essential properties. This means that in the case of nominal entities, such properties are logically necessary and collectively logically sufficient for them to qualify as the specific nominal entities in question. Nominal entities have nominal essences. However, it remains the case that we can also say of them that which sortal properties are intrinsic to the identity of a particular type of nominal object is established *a posteriori* on the basis of experience. The difference consists in the fact that the features we identify as *sortal* properties – selected from among those we ascribe to a nominal object on the basis of experience – is determined, in contradistinction to natural objects, by a theory or ultimately by one human being. It is our theories, or our intentional capacity, that determine which properties, among those experienced, attain the status of sortal properties. But – and this is my contention – once nominal entities have in this respect been defined and created, once the analytic relations have been set up, they are rendered objective and mind- or theory-independent because predicates which are true of them are true of them in every possible world. Consequently, we are also able to identify them by reference to an analytical *a posteriori* description in the same way as we do natural entities.

A couple of examples. If we consider the nature of action, we readily concede that a certain type of action – murder, for instance – is defined by certain distinctive sortal properties. That a particular murder was committed by a 25-year-old white man, using a gun, and that the victim was a 22-year-old black woman, are simply contingent features of an act, which serve to distinguish it from other generically similar acts. By contrast, to count as a murder at all a given act has to be perpetrated by one or more individuals deliberately intending to cause the death of another person, be performed in peacetime – and, to exclude euthanasia – without the victim's consent. It is accordingly a sortal feature of a murderous act that certain behaviour leads to another person's death, with the death in question being brought about by a human agent, and not caused by an animal or by an object simply impelled by the forces of nature; likewise it must hold that the behaviour was intended, and that the death was not a case of collateral damage during a military operation. It was, in the first instance, through experience that we learned that these properties are linked to actual events, and we have found them to be jointly instantiated in specific contexts of agency. In the same way, we learned that human behaviour can cause the death of others, and that such behaviour is on occasion deliberate. We have subsequently conjoined the descriptions of properties and used the resultant conjunction to define what it is to commit murder. The background to this has been an interest in guilt and punishment, and the chief focus has been the co-instantiation of these properties that make it a punishable offence. On that basis we have framed an objective determination of what conditions must be fulfilled if an individual action is to count as murder.

In the humanities we encounter defining features that do not just refer to sortal properties in types of thing, such as novels, short stories, and poems, but also in individual things such as a particular text. Recall Hans Christian Andersen's tale *The Ugly Duckling*. It is an incontrovertible fact that the story is about a duckling that grows up on a duck farm and goes through many ordeals before evolving into a beautiful white swan. This is a fact only because words and sentences are fundamental bearers of an invariant or common meaning that invests the entire text with a literal content. Through his framing of the narrative, Andersen has stipulatively endowed the duckling with a range of properties that it owns purely by virtue of that narrative. The properties in question are those of the duckling by definition and thereby become objective facts about it. It is, consequently, an identifying feature of the duckling that it emerged from the egg of a swan, and that it can think and speak. Just as we cannot talk about iron without referring to the sortal properties of iron, we cannot identify Andersen's 'ugly duckling' without noting its essential properties: that it was hatched from a swan's egg and grew up on a duck farm. Every aesthetic work is full of such facts which are just as indubitable as it is indubitable that iron rusts and can be magnetized: it is his knowledge of both the language and the everyday world that mediates Andersen's stipulation of the fictional duckling's essential properties.

It is only against the background of a literal reading that we can go on to see such facts as expressive of an allegorical meaning. It is the latter we refer to when we launch into an interpretive explanation. Why is the story about a cygnet? Why is it that the ducks and the duckling are able to talk to each other? In the process of answering these questions we offer an interpretation of the story – the traditional candidate having it that Andersen chose the duckling as an appropriate symbol of his own life, given the dramatic contrast between the grey cygnet and the white swan. The author's decision to have the cygnet grow up among ducks and hens that are perpetually bullying it means that he is thus furnished with a vehicle for the suggestion that these features mirror the course of his own life: for indeed no one took Andersen seriously, and he was jeered at for being ugly and awkward, but in the end he proves himself a talented writer. This interpretation does not exclude others, but that is a point to which we shall return.

Here, however, we can conclude that so far as facts are concerned there is nothing to support the idea that the natural sciences distinguish themselves by investigating theory-independent facts while the human sciences are quite cut off from them. Sectors of the human sciences engage with subjectivity, but the subjective unfolds through a stipulative determination of objective facts that make up the text or the work of art *qua* communicative action. They are the facts that constitute our evidence for the truth of a hypothesis, irrespective of the kind of science with which we are working. In general natural facts are explained causally, and nominal facts are explained functionally, intentionally, or intensionally, but as we shall soon see, this is no argument for claiming that there is a difference between the methods the different sciences use to articulate a hypothesis which can qualify as an explanation.

References

Kripke S. (1972), 'Naming and Necessity', in D. Davidson and G. Harmann (eds.), *Semantics of Natural Languages*, Reidel, Dordrecht.
Putnam, H. (1970), 'Is Semantics Possible?'; (1973), 'Meaning and Reference'; and (1975), 'The Meaning of 'Meaning'' (1975), in his (1975) *Mind, Language and Reality, Philosophical Papers, vol.2*. Cambridge University Press, Cambridge.

6 Methods

The opinions people commonly hold including their religious, philosophical or ethical beliefs are often not very well grounded. But in science questions are constantly being raised regarding the bases of specific assumptions. How does the scientist reach the assumptions she uses as explanations, and how can she be sure that those hypotheses are true? For if our belief that they are true is to be warranted, we require that scientific explanations be justified. And the demand made on scientific assumptions is not simply that they be *justified*. The justification has also to be reliable. This is where scientific method plays its crucial role. It must ensure that justifications are reliable. So when science provides us with explanations, they are grounded in the evidence available to the researcher, and the method ensures that the evidence is such as to make those explanations well-justified beliefs. The method lays down the rules for discovery and justification and makes scientific practice rational and intelligible.

But what methods are available to science, and what is a scientific method? There has never been agreement on the answers to these questions, but a method could arguably be characterized as a reliable rule of inference which, given a finite amount of evidence, leads us to a true hypothesis or belief. The following five types may be distinguished:

1. The inductive method
2. The abductive method
3. The inference to the best explanation
4. The hypothetico-deductive method
5. The hermeneutical method

We shall be looking at these *general* methods in the light of scientific practice. But let me first turn to other, more *specific* methods which, in one sense or another, are all bound up with prescribed rules of inference, but which also differ from them in that they serve to unify thought and action.

Rightly or wrongly, there is a tendency to speak of 'causal methods', 'quantitative methods', 'statistical methods', 'experimental methods', 'clinically controlled trials', 'double blind trials', 'interview surveys', 'qualitative methods', 'interpretive methods', 'critical methods', etc. These may in turn be subdivided into even more finely discriminated methods. There are, then, a range of specific methods which produce and are used to justify scientific assumptions and results, and the choice from among such methods is finely aligned to the kinds of object which form the focus of the inquiry. Several are clearly practical, in that they set out rules concerning how the researcher is to conduct his work in the laboratory or the study. These are *practico-instrumental* methods. They are as much rules governing thought as they are rules guiding practice. Others methods guide action to a somewhat lesser degree, but may easily be numbered among prescriptions for action. Such methods are used to ensure that the gathering of data is reliable, valid and representative. One of the epistemic goals the scientist sets himself is to procure data that may be regarded as a fair sampling of the totality of relevant data. The use of action-guiding methods ensures that the data-gathering is not random but conducted in accordance with the logic of reasoning. The scientist adheres to a set of rules which makes it possible for the epistemic objective to be achieved.

I am contending, then, that these practico-instrumental methods are to be understood as special cases of more general methods. They are basically the operationalizations of general rules of inference adapted to the domains of objects of the various disciplines. While these specific methods are linked to the individual sciences, general methods such as induction, abduction and the inference to the best explanation are common to all scientific thinking. Such methods – and the problems with which they are linked – are what establish unity among the sciences. That is why we need to focus on the question of their reliability.

Philosophers have attempted to set out the constraints to which a scientific method must be subject for it to qualify as reliable:

(i) that it be distinctive of science – that it define science;
(ii) that it explain scientific success and progress;
(iii) that it be general in its application – common to all the sciences;
(iv) that it be methodically applicable – not dependent on intuition, imagination, etc., that it may be used by any well-trained scientist; and
(v) that its reliability may be ascertained *a prior*.

The first constraint reflects the requirement that scientific method constitute each and every science. Reference to method should enable us to distinguish science from pseudo-science. Astronomy is in, astrology out.

The second constraint requires that the success of science be dependent on the method followed: appeal to method should suffice to explain scientific progress. The third constraint simply says that the method must apply without exception to every science. This means that our discussion can be restricted to general methods. Fourth, the method must be objective, independent of the person applying it. If the use of a method varied according to user it would no longer be reliable. Finally, the last of the constraints is that the validity of the method – its capability for generating reliable results – can be proved by reason. For otherwise its validity would rest on experience, and the reliability of the method would no longer be indisputable.

However, none of the methods listed above satisfies all five constraints. The hypothetico-deductive method was originally conceived of as the only method capable of meeting all five reliability constraints. But, as we shall see, the hypothetico-deductive method comes nowhere near to living up to the special status claimed for it. Moreover, it is very unlikely that science distinguishes itself from other social activities by having a method that only it uses.

Induction

Deduction is the only logically valid rule of inference – it is absolutely reliable. In using it we know that if the premises are true, so also will the conclusion be. But deduction is only applicable when we already know the premises. Unfortunately for us, we are not omniscient – only God is secure in the knowledge of all true premises, possessed independent of method. Human beings, by contrast, are limited in time and space and their immediate cognition is correspondingly constricted. We must first labour to state the premises – and premises that are true – before we can use them to ground a true conclusion. To compensate for this deficit, human beings need to devise methods by which to investigate the world, methods that are not logically valid but at best reliable.

Such methods must, in contrast to deductive inferences, be ampliative, in the sense that their conclusion states more than what the premises contain. One method is induction. This type of inference gives us at best a true hypothesis relative to our background knowledge without modifying the vocabulary that describes the available evidence. So the inductive vocabulary is constant.

Another is abduction. Abduction gives us at best a true hypothesis in relation to our background knowledge and does so through the introduction of new terms not previously figuring in our description of the world. We explain our data by postulating theoretical items not earlier entertained.

Finally, there is the inference to the best explanation, which will here be distinguished from abduction. From among a range of hypotheses this method selects the one that offers the best explanation of the available evidence. Each hypothesis will, in itself, be inductively or abductively formulated. Abduction and the inference to the best explanation are methods to which we shall return.

Returning home from work every day I expect to be able to find my way, and to find my home where it was when I left it in the morning. I can locate my address vis-à-vis the streets that I pass through in going to work and returning home. We say that it is experience that has taught me both how to get home and that my home remains in its usual spot. What has happened is that I have by this time seen my house and the roads that lead to it not once but many times, and have come to the conclusion that houses and streets do not change without there being some reason for it. If, during the course of the day, no one has demolished my house and no natural disaster has laid it, along with all the other houses, in ruins, it will still be there in the evening.

This is how we form our conceptions of the world. We infer general assumptions from individual observations. That is induction. Living from one day to the next, our store of general hypotheses about the world is forever being amplified by our current experiences. So too in science. If we want to ascertain the coefficient of expansion for copper, we begin by measuring one piece of copper, then a second, then a third, all the while modulating the temperature. When we discover that these samples expand in the same proportion each time, we conclude that all copper expands in the same way and that we have identified the coefficient of expansion for copper. Such inductive conclusions are not only useful in everyday life but are also essential to building up a body of scientific knowledge. Thus has the generality of scientific facts been established: natural constants, the orbits of the planets, chemical reactions, the causes of disease, the laws of heredity, and so on. The scientist establishes that a particular property is a constituent feature of what it is to be a particular substance, item or process. So induction is the means whereby we identify natural kinds.

The problem is, experience builds on observations made in the past. How can I know that my observations will be the same in the future? This is not merely an academic question. If I infer the general from the particular, I also infer the future from the past. Do we not often encounter examples of observations that turn out to be different from those made in the past? If we have observed white swans in the past, we cannot simply assume that we are only going to see white swans in the future. Today we know that there are black swans in Australia. Popular prejudices build on the same ungrounded inductive inferences: some Muslims are fanatics, so all are, etc. This raises the question of whether we can distinguish good induction from bad, and whether it is at all justified to claim that induction is a reliable method.

To guard against premature inferences we must impose certain constraints on the collection of the data on which an inductive inference is to be based. One important constraint is the rule of *variation, control* and *precision*. The data must be gathered with due respect to the entire reference class, every observation must be repeatable, and all the data must be gathered under the same conditions. Before concluding that all swans are white we need to have observed swans not only in Europe, but also throughout the world. The observations must be constant and independent of who makes them. Finally, they must not be influenced by local conditions of light and weather. Compliance with these constraints precludes the fallacious inference that all swans are white.

Induction is not only used to determine sortal properties but finds wide application in establishing connections between things. We do not need to put a hand on a kettle of boiling water a number of times to realize that touching a hot kettle causes a painful sensation. Once is enough. But very often the connection is not as immediately observable. To remedy this limitation, science has evolved an array of forms of statistical analysis. Such methods are simply alternative versions of inductive inference applied to data capable of quantitative expression.

Smoking causes lung cancer. But how did researchers discover that? Not all smokers contract lung cancer. Most smokers do not. So the researchers have not simply been out and about 'counting white swans'. However, it is a reasonable assumption to make that if far more smokers than non-smokers get lung cancer, then smoking is an influential factor in the contraction of the disease. The discovery of the causal nexus is based on a statistical correlation between smoking and lung cancer.

So instead of 'counting white swans' scientists divided people into two groups – smokers and non-smokers – and then compared the two with respect to the incidence of lung cancer. We may suppose that they selected

1000 smokers to be compared with 1000 non-smokers. To show that tobacco consumption increases the incidence of lung cancer, the composition of the two groups must be governed by the rule of variation, control and precision. For smoking to be a cause of lung cancer, the probability of contracting lung cancer must be greater for smokers than for non-smokers. It can be expressed thus: $P(L/S) > P(L/{\sim}S)$. But the question is, how much greater? This is where the rule of variation, control and precision comes in. We assume that it is met under a binomial distribution of the population. This means that there are just two outcomes: each person is either a smoker, S, or a non-smoker, $\sim S$. Further demands are that the probability P of finding lung cancer L is the same irrespective of who is examined, and that the chances of finding L in a given individual are the same independent of whether others also have L.

It is, however, the case with statistical correlations, as it is with simple observations, that they offer good evidence for the formulation of causal hypotheses – but not incontrovertible evidence. Even were we to discover that the number of storks in Britain always coincides with the number of births, we would not conclude that the decline in stork numbers was responsible for the declining birth rate.

Inductive inferences, then, are not logically valid inferences. So is there any possible way of showing that induction is a reliable method? This problem also goes under the name of 'the problem of induction' and was first raised by David Hume. How can we justify an inference from what has been observed to what has not been observed? Hume himself believed that it could not be justified: our confidence in induction rests on custom, and custom is a psychological disposition, which cannot be rationally grounded. But can we be satisfied with that solution? The success of science demonstrates that, in fact, inductive inferences can be made on a reliable basis. Certainly, many philosophers have looked for grounds that would justify induction.

Justification A: An inductive inference justifies its conclusion because experience shows that an argument of this form has a tendency to move from true premises to a true conclusion.

Justification B: An inductive inference justifies its conclusion because conjoining its premises to a further premise to the effect that uniformity obtains between observed and unobserved instances, leads to the conclusion.

Hume himself pointed out that justification A begs the question. That experience shows that an inductive inference was reliable in the past does not enable it to show that it will also be so in the future. Induction cannot justify induction.

The idea behind justification B is a conception of the uniformity of the nature: its contents conform to a variety of laws or rules that ensure that nothing occurs arbitrarily and at random. Granted, B makes the conclusion deductively valid, but does not this defence depend on our being able to justify the additional premise? The premise is either true *a priori* or true *a posteriori*. If it is *a priori* true it is of no interest because it tells us nothing about what makes it possible for us to have empirical knowledge. And if it is true *a posteriori*, then the argument presupposes the truth of what is to be proved. Moreover it is reasonable to assert that, generalized, the claim is false. There are many cases in which the future is different from the past. Were that not so, the world would not be a place in which wholly new things come into being and other ones cease to exist. The attempted empirical justification of induction collapses in that any such alleged justification leads either to circularity or an infinite regress.

An alternative might be to follow Peter Strawson in his bid to dissolve the problem of induction. We cannot, he contends, give a general justification of induction. However, it *is* possible for us to ground individual inductive arguments to the extent of showing that a given argument is reasonable. Strawson has in mind such examples as the reasonableness of our belief that the sun will rise tomorrow, given that it has always done so. But what does 'reasonable' mean here? If it means that this is what experience has taught hitherto then it is correct, but that we knew already. If it is Strawson's contention that it is rational to believe it, then we still need to be told why.

We appear to be left with the following question: If it is not possible to show that inductive inferences are reliable (despite our having good reason to use them) how should we be able show that inductive knowledge is possible? I would contend that we *are* in a position to do so provided we are clear about what is meant by 'a reliable method'. A method is reliable if it produces true beliefs and obviates the formation of false beliefs – not always, but as a rule. The reliability of the method is primarily determined by the reliability of the data: hence the need for the rule of variation, control and precision. Although induction is not logically valid that does not prevent it from being reliable. But that reliability is the justification. It belongs, then, to the meaning of such terms as 'reasonable' or 'substantiation' that a belief or inductive inference is justified if there is good evidence for it, whereas an assumption or inductive inference is unjustified if there is little evidence for it. And a method supported by strong evidence is reliable because, other things equal, it will generate true beliefs. In other words, it is rational to use induction even though the method cannot guarantee a veridical result. This species of reliability is the most we can hope for.

Abduction

Discoveries are not made solely on the basis of simple induction. Besides induction science makes use of another inductive method of discovery. We shall call it 'abduction'. A feature of induction is that hypotheses are formulated using the same terms as those used to describe the data. The words 'raven' and 'black' enter into our descriptions of our observations and also enter into the hypothesis that all ravens are black. Simple induction can only be conducted within the framework of one and the same conceptual scheme. Such is not the case with abduction, for it introduces and deploys fresh concepts to explain the observed phenomena.

In 1848, the Hungarian physician Ignaz Phillip Semmelweis discovered the cause of puerperal or childbed fever. He was working as an obstetrician at the Vienna General Hospital where he observed a striking difference between the hospital's two maternity wards in the numbers of women who died in childbirth: 3% in the one, 11% in the other. The only apparent difference between the two wards was the composition of the staff: midwives assisted the women in labour in the ward with the low mortality rate, while medical students worked in the ward where the rate was high.

Semmelweis made a few further observations: mortality among homebirths was at the same level as in the ward staffed by midwives; the medical students giving assistance often came straight from the autopsy room; when one of his medical colleagues died of blood poisoning after having cut himself during an autopsy he displayed the same symptoms as the women with childbed fever; and lastly, Semmelweis observed that if everyone attending the women in childbirth washed their hands in chlorinated lime before approaching them the mortality rate fell sharply – in fact, to below the level in the ward staffed by the midwives. These observations, taken together, led Semmelweis to make the assumption that it was bacteria or, as he termed it, 'cadaveric matter', that was responsible for the disease.

Semmelweis's discovery is an example of abduction. The symptoms of childbed fever may be described without any reference being made to 'cadaveric matter' (bacteria) and the cause of the condition might have proved describable using terms taken from contemporary common knowledge or medical knowledge. Indeed the conjectures advanced included ones to the effect that the cause was to be sought in atmospheric or telluric forces, in the position in which the women gave birth, overcrowding, diet – or even that it was brought on by the sight of a priest passing through the ward ringing a bell, on his way to a dying woman in the ward with low mortality – all of them drawing on conceptions current at the time. The term 'cadaveric matter', designating a microscopic agent of disease, did not figure in general medical knowledge at that time. And Semmelweis's discovery was in part derided, in part suppressed. It was not until 20 years later when Louis Pasteur isolated bacteria and Joseph Lister introduced antiseptic that medical research grasped the import of Semmelweis's hypothesis.

Induction involves a jump from the observed to the unobserved while abduction jumps from the observable to the unobservable, from the understanding of familiar to that of unfamiliar concepts. But the jump must not be uninformed if abduction is to be reliable. We have already seen that induction can be made reliable by following the rule of variation, control and precision. But what about abduction? Here many more constraints are applicable because reliable abduction must produce *relevant concepts*. An inference that introduces irrelevant concepts is not a reliable method, so abduction must be able to distinguish between relevant and irrelevant explanations. To secure the selection of relevant concepts, strict acceptability constraints must be applied to conceptualisations offered as plausible candidate explanations.

Now this would seem impossible even prior to any attempt. Surely discovery requires intuition. Abduction would hardly seem to be compatible with ideas of creativity and originality. These qualities are ones we associate with new ideas and thoughts whereas logic and method are associated with conservatism. An individual is regarded as creative and original precisely when he breaks with old norms and values. Indeed so. But even intuitive thinking observes the laws of logic, on pain of pursuing an entirely arbitrary course and seldom reaching any pertinent results.

Let us take another look at Semmelweis's discovery. There was nothing arbitrary in his lighting on the explanation that he did; indeed the discovery was directed by a string of tacit assumptions. The cause of childbed fever must, as regards time and location, coincide with the illness: attempts to identify it on the wards must be pursued along a variety of lines. The condition of women in labour in antiquity would not bear upon the investigation, nor would those in Greenland at the time. That women assisted at the birth in one ward and men on the other was not a difference relevant to the determination of the cause. (Some did in fact argue that females were more gentle than the male medical students, but Semmelweis countered that the birth itself certainly wasn't gentle.) We can call these ontological conceptions principles that concern the unity of time, place and cause.

There are ideas of an epistemological nature attaching to explanation. Importantly, the explanation must be consistent with experience and preferably also true. To achieve this aim it must be reached in a reliable way. This is accomplished in part through the framing of a hypothesis which seeks to fit with ontological principles and which is not formulated in language very different from that of prevailing scientific parlance. Semmelweis's hypothesis was able to explain the empirical differences, and to account for the observable changes that could be registered once the students had begun routinely to wash their hands. Death and decay were basic categories in medicine, and 'cadaveric matter' was a term composed of two terms already known to medicine. Semmelweis seems originally to have used the word in a literal sense to denote invisible bits of the cadavers, but later the term acquired a more metaphorical meaning. At one point Semmelweis discovered that eleven out of twelve women contracted childbed fever after having been examined by himself and his assistant. The physician and his assistant had indeed disinfected their hands in a solution of chlorinated lime before examining a woman suffering from a festering cervical cancer, but had not disinfected their hands again prior to carrying out the examinations that followed. Semmelweis thus discovered that childbed fever might be caused by contact with putrid matter originating in living organisms as well as in cadavers.

An abductive inference is drawn against the background of the knowledge the scientist already possesses and it evolves into an explanation subject to four sets of constraints:

1. Ontological principles
2. Epistemic goals
3. Methodological prescriptions
4. Semantic definitions

An explanatory hypothesis must be relevant, and compliance with these constraints should ensure that the hypothesis is relevant. We have already touched upon unity of time, place, and cause. Such principles as these are determinative of our capacity for individuating and identifying things, and stem originally from our practical ability to get our bearings. But the greater the remove from the items making up the everyday world, the more these ontological principles are violated. In the atomic world the principle of the unity of time, place and cause does not hold; in consequence, it is very difficult for us to get a grip on that world. When the focus of interest is people or opinions we will normally replace the constraint of the unity of cause by that of the unity of agency or interpretation – principles which in the same sense form the basis for the identification or individuation of an act or an interpretation. The principle of unity requires that only those things are combined that are ontologically related.

As abductive inferences can be drawn within the framework of accepted ontological principles or can violate them outright, we shall refer to such inferences as paradigmatic or trans-paradigmatic respectively. By far the most common of the two is the paradigmatic inference. If classical mechanics is our field of activity we will be concerned with paradigmatic abduction. Only when we move from theories such as classical mechanics to quantum mechanics is trans-paradigmatic abduction involved.

Characteristic of abduction is that it involves an extension of the vocabulary of science. Abductive inferences are constrained here by a number of definitions. Every inductive inference must generally be conservative in character. This entails the refusal to introduce newer concepts and superfluous concepts are eliminated – concepts that do not correspond to existing entities or properties. The method is first exposed to the constraint of semantic invariance that holds for simple induction, but if that is not achievable, a modification of terms may be attempted or, if necessary, new ones may be introduced. Modifications may affect the intensions or extensions of the terms, or both at once. We shall distinguish six modifications:

Semantic invariance: the terms in the data language and the hypothesis language have the same intension and extension.

Semantic variance: certain terms in the evidence language and the hypothesis language have the same extension but different intensions.

Semantic inclusion: certain terms in the hypothesis language have the same intension as in the evidence language but the extension of the terms is a proper subset of that of the evidence language.

Semantic separation: certain terms in the hypothesis language have the same intension as in the evidence language but different extensions.

Semantic expansion: the hypothesis language contains the data language but also includes words whose extension or intension is not exhibited in the evidence language.

Semantic correspondence: two different hypothesis languages contain the same portion of the evidence language pertinent to a specific domain.

Linguistic formulations will standardly be brought into line with accepted ontological principles. For example, abductive inferences must ultimately respect semantic variance or semantic correspondence in order to fulfil the epistemic objective of empirical adequacy and the ontological principle of the unity of cause, agency or interpretation. The hypothesis language must be so formulated vis-à-vis the evidence language that the observations can be derived logically from the hypothesis, and this is only possible if the available evidence figures as a determinative criterion for the use of expressions in the hypothesis.

Abduction is a highly complex inference. It may involve both analogies and metaphors. Philosophy of science has only just begun to apply itself to a more thorough analysis of its various forms, and it will doubtless be many years before a formal theory of abductive inferences is elaborated. Merely to identify and characterize the formal structures of analogies and metaphors is in itself a wide-ranging and severe task. But in the absence of abductive inferences science would never have seen the light of day.

Inference to the best explanation

It is standardly the case in science that choices have to be made between alternative explanations. A scientist may find that she has proposed several hypotheses on a specific subject or, alternatively, different scientists may be offering conflicting proposals as the explanation of the same phenomena. Scientists naturally want to select the hypothesis they believe to be true, and the way to achieve that aim is to single out that which offers the best explanation of the relevant data. We shall call this procedure the inference to the best explanation.

The inference to the best explanation is often confused with abduction. The difference lies in the distinction between discovery and justification. Abduction is a method for formulating a single hypothesis, while the inference to the best explanation is a method for selecting the best among the hypotheses on offer. But it might be said, and not without some, justice, that discovery and justification are two sides of the same coin, inasmuch as the inference to the best explanation is a method for uncovering the most plausible hypotheses.

The inference to the best explanation sometimes gives us the true hypothesis and sometimes fails to. But it normally ought to give us the most plausible explanation. The inference to the best explanation is an inductive inference which is not logically compelling but which we make use of constantly in our pursuit of empirical knowledge. In certain cases the alternative explanations will prove to be empirically underdetermined, or, in other words, no amount of finite evidence can show that one and only one of the hypotheses on offer is the best. We may be able to account for all of the facts we seek to have explained on the basis of diverse hypotheses even though we lack new data to make it possible to single out the one that is best. But we can never rule it out that data may exist which is explainable on one hypothesis but not on another, and that so far we have simply have failed to identify them.

A hypothesis is only the *best* given a specific norm. The norm we choose is contingent on our epistemic goals. Normally, our objective is truth, but no hypothesis can be directly compared with reality if it reaches beyond what we are immediately able to perceive. Even a simple hypothesis such as that to the effect that all ravens are black, cannot be held up against the world's ravens *in toto*. Something else must take the place of truth – something we expect to be capable of leading us to truth. So if we cannot get the best, viz. the true explanation, we must make do with the second-best, the best explanation. To achieve this we are forced to formulate certain methodological criteria which determine what counts as the best among candidate explanations. Accordingly, we believe that if these criteria are met, we have good reason to hold that the hypothesis is true. A couple of illustrations will pave the way for us in our formulation of them.

Palaeontologists agree that dinosaurs became extinct at the end of the Cretaceous Period 65 million years ago after having lived on earth for approximately 150 million years. No one knows with any certainty what caused their disappearance, and the palaeontologists are divided into two camps, each with their explanation. One camp has it that the earth was struck by a massive meteor, which then swirled enormous quantities of soil and dust up into the air. Sunlight was reduced, the temperature dropped, vegetation perished and the herbivorous dinosaurs, deprived of their sources of food, died off. With their departure the carnivorous dinosaurs died too. This explanation is supported by the fact that researchers in the Yucatan peninsula in Mexico have identified a large 300 km, meteoric crater held to date from the Cretaceous Period. But other available data fit the hypothesis too. In the fish clay that covers the cretaceous deposits, iridium has been found, a precious metal that is often found in meteorites but very rarely on earth. Here tektites, molten earth and shocked quartz, grains of sand exposed to compressive stress have also been identified. All of these phenomena are indices of meteoric impact and the possibility that it caused the subsequent extinction of dinosaurs.

Other facts are perhaps less easily explained on the basis of the hypothesis. How was it that birds and mammals were able to survive the meteor crashing to earth? Palaeontologists estimate that the birds are the direct descendants of dinosaurs, so why were these not wiped out along with the dinosaurs? Has it to do with the fact that birds and mammals are warm-blooded, whereas dinosaurs were cold-blooded? The meteor-hypothesis cannot stand on its own: it has to cohere with other hypotheses concerning some physiological factor that explains these differences.

Another hypothesis sees the reason for the disappearance of the dinosaurs in the occurrence of violent volcanic eruptions. At the very end of the Cretaceous Period, 70 to 60 million years ago, colossal volcanic eruptions occurred in the southern part of the Indian Subcontinent. Volcanic eruptions of that magnitude would have a similar effect on sunlight as the impact of a meteor, and volcanic activity would adequately explain the presence of iridium. Proponents of this theory consider it to be supported by the fact that not all dinosaurs died from one day to the next. Geological deposits attest to the fact that dinosaur numbers reduced gradually over the course of millions of years. The meteor hypothesis cannot explain this time-lag unless it can be shown that the decrease predated the fatal impact.

This example illustrates how science seeks the best explanation of a given phenomenon, but it is also obvious that agreement among researchers on which hypothesis offers the best explanation will not always obtain. Disagreement arises from deficient or inadequate data, and will only be resolved when the relevant data finally come to light (or a new hypothesis is proposed which consorts better with the existing data).

Examples of the inference to the best explanation are also met with in the social and cultural sciences. In Danish churchyards, cast iron memorial crosses are to be found dating from the beginning to the close of the nineteenth century, the period during which the first iron foundries were set up in Denmark. This fashion lasted fifty to sixty years until tombstones became popular. However, the incidence of these cast iron crosses is higher in Jutland than in other localities. An investigation into this uneven spread across the regions resulted in the framing of no less than nine distinct hypotheses, all seemingly able to explain the distribution detected. These purported explanations advanced considerations as various as access to cast iron as a substitute for wood, the size and number of the foundries, the number of deaths, how long the vogue for the crosses lasted, its pervasiveness, wind and weather, the veneration shown the dead, increased permanence of residence along with a rise in the number of family graves, and the number of burials proportional to the size of the churchyard after the highpoint of their use. What is interesting about these explanations is the fact that they are, to a considerable extent, translatable into numbers; this makes them amenable to assessment by statistical methods with a view to the identification of the best explanation. The upshot in the present instance proved to be an excellent correlation between the observations and the 'pressure' hypothesis: over the years an increase in the number of fresh burials in churchyards meant that old graves bearing cast iron crosses came under pressure. The greater the number of individuals who die in a given parish and the smaller the size of the churchyard, the more necessary it became to reuse old graves. This resulted in escalation in the rate of removal of cast iron crosses relative to the number of deaths and the size of the churchyards. It appears, then, that it is not only in natural science that hypotheses benefit from quantitative formulation, but that in the social sciences and the humanities too it is sometimes possible to put quantitative questions to the data in order to get answers involving similarly quantitative hypotheses.

The above examples illustrate the inference to the best explanation as a scientific method. On the foundation they provide we shall attempt to uncover what it is that makes one explanation best: which criteria, or rather methodological values, does a hypothesis have to satisfy to be superior to others? One value is simplicity. It tends to be difficult to specify. As a first approximation, we might say that one explanation is better than another if it does not violate obtaining ontological principles, and if it operates with only a limited number of extensions to the evidence language, *i.e.* the conceptual innovations are few and slight. However, two alternative explanations may be formulated both in accord with the same ontological principles. The issue of dinosaur extinction and that of the distribution of memorial crosses, supplied examples of this.

A hypothesis is not, then, to be assessed on the basis of one or two methodological prescriptions. There is a plurality of values to be taken into account and no hypothesis can be expected to be evaluated equally well with respect to all values. These values do not merely form the basis for the assessment of competing explanations but also play a role as methodological prescriptions for the abductive identification of relevant explanations. The list is a long one:

(a) *Precision*: a hypothesis must be precise in the sense that the implications that may be derived from it must be in demonstrable agreement with observations and experimental results.
(b) *Observational range*: a hypothesis must have the same implications as alternative hypotheses and must be able to explain all the relevant facts.
(c) *Fertility*: a hypothesis must be conducive to future research and theoretical development; it must offer a broad perspective inasmuch as its implications must be capable of reaching beyond the facts it was initially introduced to explain.
(d) *Previous success*: a hypothesis must be able to explain previous observations.
(e) *Inter-theoretical support*: a hypothesis must accord with other hypotheses and our background knowledge.
(f) *Uniformity*: a hypothesis may not be able to explain all the relevant facts but its deviation from such facts must at least be systematic.
(g) *Consistency*: a hypothesis must not be self-contradictory.
(h) *Coherence with metaphysical assumptions*: a hypothesis must be in accord with generally accepted ontological principles.
(i) *Simplicity*: a hypothesis must contain as few concepts and laws as at all possible, *i.e.* it should not introduce more entities or properties than strictly necessary.
(j) *Quantitative formulizability*: a quantitative hypothesis has an advantage over a qualitative hypothesis in that quantity increases the chances of falsification and a successful prediction more easily rules out a larger number of competing hypotheses.

(k) *New predictions*: a hypothesis must be able to predict new research results, *i.e.* it must be able to make possible the discovery of fresh phenomena or reveal new relations between familiar phenomena.

Both the meteor and the volcano hypotheses appeared to satisfy most of these prescriptions. They were both able to meet, say, (a), (c), (d), (g) and (h), but the meteor hypothesis arguably failed to meet (b) if it took the dinosaurs several million years to die out after the meteor's fall.

No hypothesis can be expected to fulfil all of these values. Alternative hypotheses will perhaps be able to exhibit values that are in part the same, in part different; and since two researchers may rank the various prescriptions differently there may not be agreement in such situations on which hypothesis is the best explanation prior to the gathering of more and increasingly precise data. In most cases the best explanation will at some point become identifiable through an assessment of the number of pertinent facts accounted for by the various hypotheses. At some future point, then, the palaeontologists too will agree on what is the best explanation of the disappearance of the dinosaurs.

The hypothetico-deductive method

Among philosophers of science, discovery and justification are often sharply distinguished even though they are really two sides of the same coin. Hans Reichenbach launched the distinction between the context of discovery and that of justification. Karl Popper went so far as to deny that there is any method of scientific discovery. Science does not arrive at new theories and assumptions by following determinate methodological rules, its hypotheses are based on conjecture, Popper avers, and stresses the desirability of bold conjectures: the bolder, the better. In consequence, we can explain discoveries only by reference to psychological and sociological factors. There are, however, methods by which belief in a scientific hypothesis may be justified. It is rational to subscribe to a hypothesis if it proves not to be falsified after repeated attempts to do so; indeed, where science is concerned it is alpha and omega that theories be repeatable and falsifiable. According to Popper, then, science operates by guessing its way to a hypothesis, which it then attempts to test scientifically, with the implications of the hypothesis being compared with the observations and experimental results of the researcher.

Popper also rejects the use of inductive inferences in science, and claims to have solved the problem of induction. But his solution is as simple as it is mistaken. It is not, Popper contends, inductive methods that are used in science but the hypothetico-deductive method. Using the logical rule of inference modus ponens, observational consequences are derived from diverse hypotheses and individual observations, and if these predictions prove to be at variance with what is open to observation an attempt is made to falsify the hypothesis by using the other rule of inference modus tollens. The hallmark of scientific practice is the use of the hypothetico-deductive method and the philosophical position that purports to justify this method Popper calls 'critical rationalism'.

Critical rationalism: Inductive inferences are not valid but empirical sciences can attain rationally based knowledge through deductive inferences and falsification.

Science, for the critical rationalist, develops as a dynamic process, in which a problem P is identified, to be explained by a hypothesis H, whose falsification is then attempted. If the hypothesis is falsified, that falsified hypothesis F gives rise to a new problem and a fresh hypothesis.

$$P_1 \to H_1 \to F_1 \to P_2 \to H_2 \to F_2 \to P_3 \to H_3.$$

Among the constraints imposed on the hypothesis is that it not be *ad hoc* in relation to its predecessor. Every hypothesis must be independently testable. No new hypothesis must be introduced simply to save that which preceded it. This constraint ensures that the auxiliary hypothesis have a testable outcome without depending on the hypothesis needing underpinning. This constraint is met if the previous theory, in conjunction with its successor, makes possible a set of fresh predictions.

Popper concedes that the Duhem-Quine thesis holds, *i.e.* the thesis to the effect that it is impossible to falsify a theory in isolation. Our assumptions about the external world never come to the 'tribunal' of sense experience individually but always as an integrated system of beliefs. When the agreement of the hypothesis with the relevant observations is tested, the conduct of the experiment is contingent on many other, often unarticulated, auxiliary hypotheses which might well prove to be false. As a consequence, a hypothesis can always be shielded from falsification if it is assumed that parts of what makes up our background knowledge are erroneous.

These constraints are illustrated by Tycho Brahe's repudiation of the Copernican system. According to Copernicus the earth orbits the sun. As the leading astronomer of his time, Tycho recognized that the earth's movement round the sun meant that the line of sight between the earth and the stars must change over a six-month period. He sought in vain to demonstrate the occurrence of such a change. Ultimately, he rejected Copernicus's theory. However, he might instead have rejected his own background assumption that the stars are relatively close to the earth. For if they are sufficiently remote, no angle will be observable by instruments available in Tycho's time. In other words, the Copernican could be able to save his hypothesis by conjoining it to a hypothesis to the effect that the stars are much further away than Tycho believed. Such an extra hypothesis would precisely not be ad hoc because its testability is independent of the heliocentric assumption.

We can distinguish, then, between two forms of falsificationism. No serious critical rationalist would endorse the first.

Simple falsificationism

(a) A scientifically acceptable hypothesis must be falsifiable, and the more falsifiable it is the better; but it must not be falsified.
(b) A falsified hypothesis must be rejected.

Sophisticated falsificationism

(a) A scientifically acceptable hypothesis must be falsifiable, and the more falsifiable it is the better, but it must not be falsified.
(b) A hypothesis is only falsified when replaceable by a new and better hypothesis.
(c) The new hypothesis must be more falsifiable than its predecessor.
(d) It must be superior to its predecessor in its ability to predict new phenomena.
(e) Falsification of a hypothesis must simultaneously be considered a confirmation of its replacement, *i.e.* the data that contradict the old hypothesis must support its successor.

We cannot prove that a hypothesis is true but we can always show that it is false. Hypotheses, though not verifiable, are confirmable, Popper holds.

A hypothesis is confirmed by the degree to which it solves problems, its degree of testability, and the rigour with which it has been tested without its thesis being affected. Note that the claim is that the hypothesis is confirmed because it could not be falsified under the most stringent conditions and not because it agrees with reality.

Naturally, falsification has a role to play in science. Hypotheses that do not accord with observations and experiments are often rejected. But there is no reason to hold, with Popper, that scientific practice reflects the hypothetico-deductive method, and even less that critical rationalism can justify the hypothetico-deductive method.

No scientists guess their way to their hypotheses. None sets up hypotheses just to knock them down. Hypotheses are framed, rather, to explain phenomena that, without explanation, stand unexplained and incomprehensible. If it proves the case that the hypothesis fails to meet the requirements we make of a good explanation – including its being able to explain all the data it is designed to explain – then the researcher may reject it as an inferior hypothesis. In many cases a hypothesis will be able to explain some data but not others that it was expected that it would. But this does not mean that it is automatically rejected. Even if a hypothesis appears to have been falsified outright, there remain many other assumptions that the trained researcher will seek to modify.

Ever since the Greeks, scientists had believed that the earth remained stationary at the centre of the universe. And when Copernicus advanced his hypothesis to the effect that the earth turned on its axis and orbited the sun, there were indeed a number of observations fully congruent with the hypothesis, but there were others that appeared directly to contradict it. All the same, his learned contemporaries did not reject the heliocentric system out of hand. Rather, the astronomers sought to understand why there remained certain intractable phenomena that the hypothesis seemed unable to explain.

Very few hypotheses are directly applicable to what the researcher observes. Especially in light of the hi-tech profile of the natural sciences of today, scientific experiments involve a wide range of electronic instruments and computers for the processing of data, with the result that the sources of error are multiple and unpredictable. Today, natural science no longer appeals to observable phenomena in support of its assumptions. Data have replaced phenomena, *i.e.* instrumentally obtained information has replaced the observation of phenomena by human sense organs: thus, there has been a marked shift from qualitative to quantitative evidence, and phenomena are now such evidence.

Scientific data are the result of a large number of theoretical hypotheses and the processing of a vast amount of experimental information. It has become ever clearer that our observations are theory-laden. We would not be able to perform the observations that are possible today had it not been for our elaboration of complex theories of the workings of instruments, of their efficiency, reliability and theories of signal reinforcement, noise filterability and data processing. Negative data always raise the question as to whether our understanding of these theories is in fact inadequate, whether the actual experiments are well understood or technically problematic, or whether the hypothesis tested should be rejected. Scientific data are interpreted phenomena that fail to falsify a hypothesis if these phenomena are given a different interpretation. Such factors do not sit comfortably with the claim that natural science is conducted according to the hypothetico-deductive method.

Furthermore the falsification of statistical hypotheses is not easily achieved. Contemporary science reaches its results not least by working with probabilities. This proceeds in two ways. Either, the scientific hypothesis itself is formulated in terms of a statistical proposition, or the investigations which ground a scientific hypothesis that does not itself take, say, the form of a statistical proposition, rely on a statistical analysis of the recorded data.

An example of a statistical hypothesis is the claim that smoking causes cancer. Since not all smokers get lung cancer the hypothesis cannot be framed as a universal law that all smokers do. Instead, the causal link between smoking and cancer is expressed as a relation between probabilities. We can formulate a hypothesis to the effect that the probability that smokers develop lung cancer is higher than that for non-smokers. The probability is about ten % for smokers and around one % for non-smokers.

It is not difficult to see that it is almost impossible to falsify such a claim. If the hypothesis is to hold it must be anticipated that 10 out of a hundred smokers will contract lung cancer. To falsify it, it must be possible to show that only one does. But how is that to be achieved? If we examine the first hundred smokers we meet in the street and find that none of them develop the dire disease we will not have refuted the claim that smoking is one of its causes. It may be wholly fortuitous that none of these people develop cancer or it may just be that those we run into are all members of 'Senior Citizen Lifelong Smokers' – five score nonagenarian smokers on their annual outing. Before doubt is cast on the alleged link between smoking and cancer, the number of those surveyed will need to be significantly extended; subjects need to be randomly selected, and to constitute a representative spread of smokers in terms of gender, age, occupation, nationality etc. But from a logical perspective even that is not enough.

In principle, no statistical hypothesis can be falsified on the basis of a finite quantity of data. The probability of throwing a six with a die is normally 1/6. But if one throws 12 sixes on the first 12 throws of a particular die, it has not been disproved that one's chance of throwing a six is 1/6, even when using that same die. It is conceivable that not one of the next 72 throws will be a six. So rather than trying to falsify a hypothesis, we can instead formulate the conditions under which we would not consider it confirmed – which is not the same as to falsify it. Such an aim is not objective; it builds on the conception of an appropriate statistical relationship between the hypothesis and the data it is presumed to support, and what counts as 'appropriate' is determined by a norm. Only when the correlation obtains can the hypothesis be said to have received confirmation.

Popper, as we have seen, rejects induction. The cornerstone of the hypothetico-deductive method is falsification. Unfortunately for Popper, critical rationalism is built on sand. He is right in thinking that falsification is logically compelling – if we disregard the Quine-Duhem hypothesis – as contrasted with verification. But the point is that there is no reason to take falsification seriously unless one believes that the future will resemble the past. If the future did not resemble the past, why should the falsification of a hypothesis lead to its rejection? The bar of falsification must be equally able to figure as such tomorrow. Only if credence is given to induction is there any need to take the negative result seriously and act accordingly. Falsification presupposes induction if it is to determine of agency in scientific practice. Thus, Popper has not solved the problem of induction, and he has failed to show that the hypothetico-deductive method is what we encounter in science.

The hermeneutical method

A general method is a reliable means by which thinking can obtain a possible explanation. If thought follows the rules specified by the method, it will, more often than not, be able to terminate in the framing of a relevant hypothesis. Such a method is formal in the sense that it works irrespective of the content of thought. What ensures the method's reliability is its form, not its materials. This leads to the expectation that such methods will not find application only in natural science but will do so equally in the social and human sciences.

This notwithstanding, many scholars and social scientists have thought that the fundamental differences between nature, human beings and society imply crucial differences in how they are to be studied. Indeed, it has not even seemed necessary to argue for that view. Anyone can quite literally see the difference between the experiments of physics and picture analysis of art history. The difference consists, it is said, in the fact that natural science uses the inductive method – or possibly the hypothetico-deductive method – while the humanities and the social sciences make their inferences by drawing on the hermeneutical and the so-called 'critical method'. Thus Jürgen Habermas contends that scientific method reflects our epistemological interests: these fall into three categories, which are sharply defined in relation to each other and mutually incompatible. Natural science satisfies human technological interests and does so through its employment of the empirico-causal method; the cultural sciences are founded in hermeneutics, *i.e.* a non-nomothetic, non-objective method grounded in a practical interest in understanding and empathy; and lastly the social sciences work on the basis of the critical method, where interest may be focused not only on identifying social laws, but also to an equal degree on the exposure of ideological manipulation, the illicit exercise of power, and social oppression. This claim, however, has never seriously been put to the test and it proves untenable as soon as the worst ambiguities and misconceptions have been removed.

The hermeneutical method does indeed, at first glance, differ from other experimental methods. But I shall be arguing that hermeneutics, on a par with experiment, belongs to the specific practico-instrumental methods that a researcher may draw upon, depending on what material she intends to study and her choice of epistemic goals. Habermas's talk of epistemological interests is of course relevant to any analysis of scientific explanation, as I argued in the previous chapter, but these are determined by diverse ontologies and not by incompatible methodologies. The experimental method is the natural choice of the researcher pursuing causal explanations, but she can have recourse to hermeneutics or the critical method, if it is intentional or interpretive explanations she is after. However, the methodology is basically the same in each case in spite of different epistemological interest.

Hermeneutics is normally defined as a method that leads to understanding. This definition is not, however, exhaustive. Were that so, hermeneutics might also have been used to mediate other explanations since all species of explanation could potentially contribute to an increase in understanding. It is more apt to claim that hermeneutics is a method that leads to an understanding of meaning. This applies both in the case of the meaning that attaches to individual actions and that bound up with the social outputs delivered by such actions. The latter include linguistic expression, paintings, sculptures, musical compositions, social activities and institutions.

For human actions are indeed meaningful. Behaviour becomes agency if directed by desires or aims. Many actions in a social context do themselves create meaningful outputs. A civil law is meaningful but a hammer is not. A tool merely has a function; a civil law has both a functional and a representational element. Such a law focuses on particular kinds of action, and such focus invests it with social meaning. The differences can briefly be described as follows:

1. Social actions have meaning. There is a difference between human *action* and animal *behaviour*.
2. There is a difference between what an action 'means' and the significance the agent herself confers on the action. The difference is one between *sign* and *symbol*.
3. Social actions, as against animal behaviour, are shot through with normative conventions. The difference is one between *habits* and *practices*.
4. The higher animals may well have beliefs that direct their behaviour, but human actions are governed by theories about the world. The difference is that between *regularities* and *norms*.

Many will hold that a scientific understanding of people and society must begin from such ontological differences and be prosecuted using the hermeneutical method, which takes account of the intentionality and meaningfulness of actions. Indeed, the hermeneutical imperative has it that actions must be understood from the inside, *i.e.* as rational and rule-governed conduct viewed in relation to the agent's own experience of the world.

Nonetheless there are other perspectives on society that deliberately disregard human agency both on the individual and the collective plane. Social science disciplines such as economics and political science seek to comprehend social institutions and organizations as ontological entities not involving human properties such as intentions, willing, aims, desires, rule-following, etc.; they are seen as entities that may be described causally in terms of economical and social forces or structurally by using game and systems theory. While some social scientists contend that a description of social forces and institutions cannot be reduced to a description of individual actions, others seek to understand society precisely on the basis of such a description. Ontology determines our epistemic goals and thus the content of our hypotheses, but the ways in which we establish and assess these hypotheses are not incommensurable.

There are no methodical oppositions between natural science on the one hand, and the human and social sciences on the other, even if it is insisted that hermeneutics can and should be applied in the social and human sciences. Dagfinn Føllesdal makes the same point when he characterizes hermeneutics as the hypothetico-deductive method applied to meaningful material. However, this characterization is not the most felicitous inasmuch as Føllesdal's conception of the scientific method as hypothetico-deductive is the expression of an outmoded idea. What is right in Føllesdal's analysis is the notion that hermeneutics does not signal a radical methodological difference between the human sciences and the natural sciences but reflects, rather, their ontological distinctions. Hermeneutics is determined by its materials and by the epistemic goals that we set for the human and social sciences. It does not constitute a different methodological form of inference that has its own logical principles. Rather, hermeneutics is the application to meaningful material of the same inductive and abductive methods as well as the inference to the best explanation as are applied in causal nomothetic sciences. The difference lies in the fact that the proposed hypotheses are not causal explanations but structural, functional, intentional or interpretive explanations.

Whenever we read a text or study a work of art, we are constantly in the business of forming a provisional hypothesis about its content on the basis of inductive or abductive considerations, a hypothesis subject to ongoing revision as our acquaintance with the sources, text or artwork deepens. A text presents us with a host of facts that lead us to infer a particular hypothesis concerning parts of the text or all of it. This applies irrespective of whether the text deals with historical episodes or has a fictional content.

The method can be illustrated by an example drawn from literary studies. Føllesdal describes how the Stranger in Henrik Ibsen's play, *Peer Gynt* has been variously interpreted as Anxiety, Death, Ibsen himself, the Devil or Lord Byron. This singular character appears suddenly, standing next to Peer in darkness on the deck of the ship and reappears in a later, episode that finds Peer, after the shipwreck, sitting astride an upturned boat. Ibsen specifies a number of facts about the stranger, which provide the starting point for any pertinent interpretation. The stranger is a man. He is white as a sheet, swims using his left leg, relishes storms and shipwreck, takes a scientific interest in anatomy, wants to be a moral guide for Peer, comforts Peer by saying that no one dies in the middle of the fifth act, can float if he just holds on to a crack, etc. Other facts concern Peer's relation to the stranger. He regards him as a freethinker, as a tedious moralist and says to him: 'Get out of here!' and 'Get out of here, scarecrow!' The scholar engaged in literary interpretation uses these features to form hypotheses as to what the stranger represents. But she will also draw on other data embedded in her background knowledge. This might include general knowledge about the nature of anxiety, the author, the devil, Byron or the period in which the work was penned. In the course of reading the play the interpreter familiarizes herself with the role of the stranger; she focuses her attention on certain passages that suggest to her that the stranger might be the devil. The words 'Get out of here!' remind her of Jesus' words to the devil, and the fact that the stranger swims using his left leg is consistent with the fact that the devil is usually portrayed as having a hoof on his right foot, etc. Next, the interpreter must check whether the hypothesis is consistent with other passages. An interpretation must not just conform to the facts supplied about the stranger but also to the work as a whole, which is to say, to all the circumstances engendered by the work.

This process is known as *the hermeneutical circle*. We need to be able to understand part of the work, for example, the character of the Stranger, in the light of the whole, and the whole in the light of the parts. In other words, theoretical coherence must obtain between the interpretation of the parts and that of the work as a whole. If the interpretation of the stranger fails to fit with the whole, the proffered hypothesis loses much of its credibility.

It is in hermeneutics too that the concept *horizon of understanding* is encountered. This refers to the fact that both interpreter and object of interpretation are bearers of a wealth of explicit and implicit beliefs and attitudes that are simply taken for granted. The horizon of understanding may be said to comprise our norms, values, prejudices and background knowledge – all of it entering into a unified world-view. Much recent hermeneutics is concerned with an insistence that it is impossible for us to shed our horizon of understanding. Any interpretation is underpinned by particular beliefs and attitudes which – were they to be analysed – would show themselves to repose on still other assumptions. We may have the capacity to become aware of parts of our horizon of understanding but never of the whole at once so as to possess a grasp of all the preconditions on which an interpretation proceeds. The interpretation of a work will always depend on the interpreter's horizon of understanding, which will vary from one interpreter to another. This shared understanding is what Hans-Georg Gadamer called 'the fusion of horizons'. Optimally, the understandings of the interpreter and the author fuse, and in the worst case the differences obstruct the path to a collective consensus.

Shared understanding is to be achieved through the use of a reliable method. The success of the method is contingent on its reliability, and its reliability contingent on its capacity to generate a fusion of horizons – that is, to produce consensus across differences in values, prejudices and background knowledge.

What are the implications of the horizon of understanding for hermeneutics? Is the hermeneutical method reliable? And how can we be sure that the method delivers a relevant interpretive explanation? Underlying that question is the fact that we normally distinguish between interpretation, over-interpretation, and misinterpretation. Neither over-interpretations nor misinterpretations are relevant explanations, and to the extent that hermeneutics is hospitable to such it fails to offer a reliable method.

Postmodern and deconstructivist critics regularly argue that in putting questions to the text, the interpreter need not feel bound by it. There is no such thing as an over-interpretation – all interpretations are of equal worth. This view is nothing less than scientific charlatanism. Inconsistent hypotheses, for example, are scientifically unacceptable because any statement whatever can be derived from a self-contradictory interpretation. And to avoid inconsistent explanations, empirical adequacy and coherence must at the very least be included among the epistemic goals we require of any interpretation that it meet.

Now these critics will roundly deny that a serviceable concept of empirical adequacy can be distilled from a reading of the texts since one and the same text offers disparate sets of data: the way in which sign becomes symbol is, in part, a product of interpretation. Different interpreters are going to propose different interpretations because the data they invoke in making their separate cases are the facts *as seen* from the perspective of that very interpretation. We are caught in the hermeneutical circle. But we have been down this road before – in the argument for the theory-ladenness of observation.

I would venture the claim that the notion of theory-ladenness is overhyped in the philosophy of science. Such overemphasis forgets that nature itself evidences sortal facts that figure as objective determinations of the objects in the domain of natural science. By the same token, in my view, the theory-ladenness of interpretation is disproportionately highlighted in the human sciences. We must distinguish between the reliability of data and method. Just as abduction can produce empirical underdetermination in natural science, it can produce like underdetermination in the social and human sciences, which – insofar as it concerns intentional and interpretive explanations – we might call *behavioural* and *intensional* underdetermination. But alternative hypotheses are underdetermined in relation to the *same data*.

Every interpretation must begin from a literal reading of the text that is invariant from one reader to another in virtue of the fact that the words receive their meaning from their use in accordance with the rules of a given linguistic community. As Ludvig Wittgenstein famously contended, there is no such thing as a private language. That language use is, from one language user to another, a constant, means that we are able to use it to communicate unambiguously. In the general run of things, we understand what we read, and in recounting that content to others we make ourselves understood without recourse to interpretation. So without a shared linguistic understanding of the fictional facts contrived by the text there would be no interpretation of the non-literal aspects of the text on which agreement is sought. If no such understanding is possible, the text could hardly purport to appeal to the reader's understanding, because the interpretation itself would have no literal sense, and author and interpreter – interpreter and interpreter – would be sealed off from each other.

There *is* literal meaning, then – meaning which is not the product of interpretation. We set about interpreting when we fail to understand. We proceed in a reliable manner from uninterpreted understanding to interpretive reading by means of abductive inference. This involves the imposition of both internal and external constraints. *Internal* principles include – certainly until the twentieth century's jettisoning of it – the constraint of the unity of time, place and action. An interpretation must so view the text that the story, or the sequence of events, is presented as cohesive – temporally and relationally – by the causal threads of the narrative. Nothing in the text is arbitrary. Each part, each scene, each action must figure as consequent upon preceding events and as a cause of those that follow. Figuring in such a sequence are the characters' thoughts, feelings and desires – manifest as well as hidden – as causally operative components in an action-creating pattern. Each item is a constituent in an overall narrative informed with a causal structure. Among the *external* principles that are relevant are linguistic rules and norms, authorial intentions, the *Zeitgeist* and culture. A primary constraint is that a reading should be admissible under the public linguistic norms that govern time and language in which the text is written. The originator of the work wished to convey something through it; it is marked by her experiences, her command of the resources of language and of the times in which it is written. And the work has a history. All these features are present as ontological preconditions of any understanding of the work. A plausible interpretation will thus be both limited by, and subject to, these determinations. Whether or not one believes that authorial intentions should be drawn into the interpretation will depend on one's epistemic goals. But we shall defer that question to Chapter 9.

For the moment, let us turn our attention to the methodological prescriptions that in fact shape the articulation of interpretations and the inference to the best of them. Even though the phrasing is different they are in fact the same values as previously:

(a) *Precision*: an interpretation must be in close agreement with the explicit features of the text selected as evidence for an abductive inference.
(b) *Observational range*: an interpretation must not only be able to explain the same facts as other interpretations but must be able to explain all the relevant facts about the text.
(c) *Fertility*: an interpretation must be able to explain other concrete aspects of the text that were not its primary targets and must, optimally, be conducive to future research and theoretical development.
(d) *Previous success*: an interpretation must be able to explain earlier readings.
(e) *Inter-theoretical support*: an interpretation must not contradict other explicit features of the text or our background knowledge; instead it must conduce to semantic coherence between the individual parts of the text.

(f) *Consistency*: an interpretation must not be self-contradictory.
(g) *Coherent with metaphysical assumptions*: an interpretation must agree with the internal and external ontological principles of the text.
(h) *Simplicity*: an interpretation must use few and simple concepts and linking elements; it must not introduce more meanings than the context requires.
(i) *New predictions*: an interpretation must make new readings possible and open up for new observations or readings of the text.

The constraints of uniformity and quantitative formulation patently cannot be imposed on hypotheses concerning reference and representation since, by virtue of their nature, these can only be expressed in qualitative language. Meaning and views of life cannot be measured and weighed.

These methodological principles appear simple in comparison with the highly complex interpretations that the reading of an entire work may prompt – *too* simple, some will argue. And it is true that it is possible to specify more precisely the content of these prescriptions only when they are applied to particular texts. But it is worth remembering that the overall interpretation of a work consists of a myriad of partial interpretations, with each hypothesis being the product of many individual observations. Methodological prescriptions are not explicitly evoked to in the process of discovery; researchers do not go around reminding themselves or each other that they should observe them. What they do concentrate on is the resolution of a problem of understanding – not on the way in which that is achieved. Nor do we keep recalling to mind the rules of the road when in traffic or moral rules when in contact with others. But we normally observe them all the same. Methodology is implicitly embedded in our reasoning when we embark upon the interpretation a text. It underpins hermeneutical practice. But only when the philosopher sets about describing this practice do the relevant prescriptions come to light.

What, then, is the answer to over-interpretation and misinterpretation? In light of the inference to the best explanation we may construe over-interpretation as a hypothesis that offends against the simplicity constraint. It violates the principle of economy and introduces a superfluity of meaning into the interpretation that is plausibly at odds with ontological principles, since it flouts the internal and external constraints imposed on the work. It is a constraint on any interpretation that – with due regard to the ontological principles – it seeks to use only those concepts, analogies, and representations which are needed for understanding the textual content.

Given two explanations, A and B, which interpret the same data, B will be an over-interpretation in relation to A, if A achieves its objective using fewer means than B. Misinterpretation, by contrast, offends against the constraints of precision and inter-theoretical support. A lack of fit between the hypothesis and the evidence marshalled in its support will be apparent. The inference to the best explanation disallows over-interpretations and misinterpretations, but it does not eliminate those interpretations that are intensionally underdetermined by the text. The reliability of the method does not exclude various interpretations that prove equally plausible.

Hermeneutics is a scientific method in that it is an instance of the same forms of discovery and justification within those studies whose object is meaning, as the experimental and statistical methods are within natural science. However, interpretation does not admit of closure: fresh theories point the way to new interpretations, and fresh aesthetic, ethical and cognitive values add further perspectives to what we already understand. Whatever difference there might be between natural science and human science on this point turns on the subject matter and not the method. For we meet open-ended interpretation outside the human sciences too. Interpretations in natural science go through transmutations *pari passu* with the redefinition of epistemic goals and the acquisition of fresh knowledge. While the Copenhagen interpretation of quantum mechanics was in the ascendancy among physicists up to the end of the sixties, it is today in contention with other theories. Indeed, as we shall see, not only interpretations in natural science but also causal explanations change with our desires and interests.

References

Føllesdal, D. (1979), 'Hermeneutics and the Hypothetico-deductive Method', in *Dialectica*, 33, 319-336.
Habermas, J. (1971), *Erkenntnis und Interesse*. Frankfurt-am-Main: Suhrkamp 1971.
Hendricks, V.F. & Faye, J. (1999), 'Abducting Explanation,' in L. Magnani, N. J. Nersessian, and P. Thagard (eds.) *Model-Based Reasoning in Scientific Discovery*, Kluwer Academic/Plenum Press, New York, 271-93.
Hutchinson, London.
Newton-Smith, W. (1987), *The Rationality of Science*. Routledge & Kegan Paul, London.
Popper, K.R. (1959), *The Logic of Scientific Discovery*.

7 Laws and Rules

Newton discovered that the force that causes the apple to fall to the ground is the same as that, which keeps the moon in its orbit of the earth, governs the tides and causes the swing of the pendulum. Newton sought to uncover the all-encompassing unity in nature's manifold. The literary scholar who interprets Molly's dream in James Joyce's *Ulysses*, or the historian who traces Napoleon's rise to his inauguration as Emperor of the French, seeks to understand the actual text or an individual sequence of events. The scholar is not interested in whether Molly's dream is at all like Nebuchadnezzar's or his own dreams. Nor is the historian interested in knowing about Caesar's ascent to high office – such information will tell him nothing about the friendships, alliances, favours and political currents which propelled Napoleon to the pinnacle of power.

Natural science and human science appear to be concerned with two different cultures, with the role of the natural scientist being to provide nomothetic explanations and that of the scholar to supply idiographic descriptions. Natural science uncovers nature's laws, *i.e.* it articulates universal propositions that specify how a type of phenomena behaves in conjunction with other types of phenomena. The aim, then, is to give an account of the commonly occurring and typical, and not to describe the world in all its individuality and variety. If, however, we shift the spotlight to the humanities we encounter instead the project of understanding phenomena *qua* individual and singular events. Here significance accrues to the individual item through its being set in a particular context, which may be historical, literary, social or cultural; the item's full import can only be understood when the relevant context and individual values enter into the interpretation. In consequence, epistemic goals such as truth assume different forms according to whether they take the form of regularities in natural science, or singular descriptions in human science, and the methods deployed within the two subject areas will also be different. So much for the received view.

The alleged gulf between the two cultures reposes, in my view, on a myth, and it is a myth needing to be disposed of. There are patently individual events which are the focus of natural science research: they include the Big Bang, Mars's orbit of the sun, the sun's magnetic field, the evolution of the earth, the extinction of dinosaurs, the ozone hole over Antarctica, the glacier brooks in Greenland, etc., even as there is a wealth of events that form focuses of research in the humanities and are studied with respect to their general and typical features. Only consider the burial rites of the Bronze Age, English phonemics, the thematic content of ballads, the teaching of foreign languages to the children of immigrants.

In justification of the distinction one might appeal to the fact that physics, chemistry and biology concern themselves with natural objects that exist prior to any description being given of them, simply because the world contains clearly bounded and naturally occurring kinds whose members are distinguished by their having certain properties in common. The nomothetic features of the explanations of natural science result from the description of the connections between the different kinds. Natural laws are simply the causal connections between the diversity of types occurring in nature. In the human sciences, by contrast, there are no naturally occurring kinds. There is nothing there that corresponds to atoms, molecules, tigers and elephants. Such domains are occupied by mental, social and cultural constructions. Genres and styles are the creation of human subjects. In consequence, there are no laws to be identified within the human sciences.

To my mind such a distinction between natural science and the humanities is superficial and specious. Before seeking to establish whether there are laws in the human sciences we must know what a law is. All too often the concept of law is denied a place in the human sciences without any clear conception of what the notion involves. It appears, for instance, to be forgotten that the description of a natural kind also comes about through abstraction and idealization. When a researcher seeks to offer a scientifically applicable characterization of a natural kind, she focuses on certain properties while ignoring others featured by all the members of the kind under scrutiny. In this discrimination, then, there is no significant difference between the natural and the human sciences. And even though it is correct to say that the subject matter of the human and social sciences are products of a construction, it is also true that we have so constructed them that they form kinds whose members evince one and the same set of properties. There may well be disagreement on where the boundaries should be drawn, but this does not preclude the typical members evidencing a high degree of uniformity and invariance.

In the present chapter we shall be looking at how the distinction between the universality of explanation and the individuality of interpretation easily belies the nature of the subject matter of the sciences and their practices. For natural science also offers explanations of individual and concrete phenomena. It is the behaviour and development of particular systems that science explains. Hempel and Oppenheim's explanatory model – for all its deficiencies – reveals the individuality of explanation with all the clarity that could be desired, and it is cause for wonder that so many have overlooked this fact. Philosophers of science have focused mainly on the presence of nomic premises in explanation. Given the structure of models, we can only explain the explanandum if the premises of the explanans include propositions about the actual individual system, *i.e.* propositions about the so-called initial conditions. We can never offer a general explanation. When we use Newtonian mechanics to explain the swing of the pendulum, the rise and fall of the tide, or particular planetary motions, it is some specific pendulum at a particular location on earth, the tide in a particular locality, a particular planet's motion, which is explained. It is impossible for us at a stroke to account for all pendulum movement or all planetary motion because nowhere are Newtonian laws unqualifiedly in evidence. No phenomena conform to these laws exactly.

Conversely, reference to a context is no guarantee of unique interpretation. Every context must necessarily be described in terms of concepts that, if apposite, may also be applied in other contexts, and so no interpretation can, strictly speaking, be said to be unique. Shared rules and conventions make actions homogeneous and generally understandable. When we produce artefacts, language, texts, paintings, we do so with the aim of using them as mediums of expression, and our being understood is contingent upon our adherence to recognizable rubrics and standards. Indeed, contextual dependence is not restricted to the human and social sciences: we shall see how it figures in natural science too in the form of *ceteris paribus* clauses and epistemic interests.

The minimalist position

Our belief that laws exist in nature can be traced back to human conceptions of a divine creator of the universe. God ordained the laws to which everything in nature is subject, even as the king decrees the laws the members of society must obey. The laws of society regulate the activities of citizens, and it was thought that the laws of nature operate analogously. When science abandoned the idea of a creative deity it did not renounce the idea of a law of nature. For even though God no longer figured, the factors that had prompted us to believe in the laws of nature remained. Nature evidences a high degree of regularity. Many things follow invariant recurrent patterns. All water boils at a hundred degrees Celsius, all gold dissolves in *aqua regia*, and all humans die before reaching their 150th year. These appear to be three different laws of nature.

Each of these laws of nature would appear to be renderable as follows: it is a law that all *F*s are *G*s where '*F*' and '*G*' stand for properties of one or more things. For instance we might say that that all matter that has the property of being water also has the property of boiling at a hundred degrees.

We shall now look at three traditional proposals as to how a law of nature should be understood, and I shall seek to show that none of them offers a satisfactory analysis. As an alternative I shall present a fourth possibility.

Let us begin by fixing ideas: We distinguish laws of nature from nomic propositions and theories of laws. Naturally, nature's laws are not identical with their linguistic or mathematical representation. We also believe that we can establish that laws exist on the basis of our observation of a regular connection between different phenomena. Many laws seem to be directly observable. But the question is whether laws are other and more than what it is possible for us to observe. Do the laws of nature subsist as existent entities or structures underlying the appearances of things, or are they merely a manifestation of the regularities we find in nature? What is the nature of the laws of nature? This is the question which any philosophical theory of the laws of nature has to answer.

We arrive at an empirical generalization when we are able to establish that all observed *F*s are *G*s. But we cannot immediately infer from that, that all *F*s are *G*s. A law cannot be identified merely on the basis of what we have observed. But what else is needed? Could we say that a law is identified on the basis of what we could, in principle, observe, if we just had the time and the means?

Some philosophers would reply in the affirmative. This view goes back to David Hume. The adherents of such a *minimalist* theory lay it down that a law of nature is simply a regularity obtaining between *types* of phenomena, and that we become come to believe such regularities will continue in the future by our use of induction. A law of nature is, then, a uniform pattern obtaining between all instances of phenomena of different types. If we were able to trace every single token of these types, Fs and Gs – in the past, present and future, and throughout the universe – we would observe that all Fs are Gs; but this regularity – that all Fs are Gs – is all there is to a correct conception of law.

Definition: It is a law that Fs are Gs iff (read: if and only if) all Fs are Gs.

This definition commonly meets with a battery of objections aiming to show that the concept of law is in fact much richer in content.

First and foremost it is contended that regularity is not *sufficient* for a law. The reasoning is simple: Assume that we investigate all gold and all uranium in the universe and on completing our investigation are able to conclude that: (a) All existing lumps of pure gold-195 possess a mass of less than 1000 kg; and (b) all existing lumps of pure uranium-235 possess a mass of less than 1000 kg. Both (a) and (b) are true generalizations but only (b) expresses a law, owing to the fact that uranium-235 has a critical mass of less than 1 kg and therefore a lump of over 1 kg would explode. There are no physical limitations of this kind governing lumps of gold of 1000 kg. (a) refers to an accidental regularity while (b) refers to a nomic regularity.

A characterization of laws of nature is the same as an identification of the concept 'law of nature'. It is important that any such account of them makes clear that it is the laws themselves that are so designated and not their linguistic or mathematical representations. Our philosophical intuitions, going back to the notion of God as lawmaker, dictate that to qualify as such, a law of nature must display a certain set of properties:

1. A law of nature is universal; it must apply to all relevant cases and not be limited in scope.
2. A law of nature is explanatory, or, more precisely, the relevant phenomenon must be explainable by reference to the law.
3. A law of nature holds of necessity.
4. A law of nature must satisfy the counterfactual principle.

A modern philosophical analysis must seek either to explain these intuitions or to replace them by others.

In the first place we distinguish between empirical generalizations and laws. If we can presently establish that at this point in time all Britons are less than seven foot five inches in height, no one would claim that a law of nature has been discovered. What we would have is an empirical generalization. At some other point in time there might well be a Briton whose height exceeded seven foot five inches. A law is not limited by time and place. And even if there never was, nor will ever be, a Briton towering above the seven foot five inches mark, that fact would still not be regarded as a law. The mere reference to 'Britons' limits the validity of the generalization as a law.

But the British philosopher Frank Ramsey has drawn attention to the fact that an empirical generalization can always be so formulated that its validity is not limited in time and place. Instead of the formulation, 'All Britons now alive are under seven foot five in height' we might have said: 'All inhabitants of a country with a population of 58 million, and whose capital is called 'London' and boasts 7 million inhabitants, and whose head of state is a Queen, mother of four offspring, and which is governed by what is termed a 'Labour' administration, are under seven foot five in height.' Such a sentence expresses a universal proposition that applies to anything satisfying the properties mentioned in it. Its validity is not restricted to the year 2001 or to the United Kingdom. Universality is obviously a necessary but not a sufficient condition of lawfulness. We need then to proceed further in our investigation of the other conditions.

The next condition demands that reference to a law must be able to figure in an explanation. The fact that all Britons are less than seven foot five in height offers no explanation at all as to why the sitting British Prime Minister is less than seven foot five. A contingent fact cannot explain another contingent fact, and all empirical generalizations are contingent. Furthermore, the British Prime Minister belongs to the set of all Britons, so his height cannot be explained by reference to the height of Britons. Nothing can be an explanation of itself.

This argument can be resisted if we convert the contingent generalization to a Ramsey-generalization. If it is fully legitimate to explain electric charge conservation in a concrete physical decay by reference to the fact that all electric charges are conserved, it would seem equally legitimate to explain the height of the current British Prime Minister by reference to a Ramsey-generalization. The claim that there is a difference requires an argument to the conclusion that the laws of conservation are more than mere Ramsey-generalizations. Explanatory power is, in consequence, also only necessary, not sufficient, for lawfulness.

Third, a law is thought to hold with necessity. Assume that it is a law that all Britons are less than seven foot five tall. This obviously entails that any British Prime Minister necessarily will be less than seven foot five tall.

An empirical generalization cannot impose such a necessary property on British Prime Ministers. There are people over seven foot five tall. So it cannot be a law.

Here too the Ramsey generalization fulfils its objective. There are no cases falling under this formulation that do not necessarily satisfy the specification that the law expresses. Of course there could easily be holders of the office of British Prime Minister over seven foot five in height but the present incumbent only satisfies the Ramsey generalization because, in common with Prime Ministers in infinitely many possible countries, he fits the description. The Ramsey-generalization does not pronounce on all British Prime Ministers. Again, the condition is only necessary.

Fourth, we have the counterfactual constraint. Assume once more that it is a law that all UK citizens are less than seven foot five tall. Find an American seven foot seven inches tall. It must follow that had he been a British Prime Minister he would have been less than seven foot five tall. But make him Prime Minister of Britain and it will be discovered that he is still seven foot seven. So an empirical generalization cannot be a law.

In the face of this constraint Ramsey's proposal appears impotent. For any possible x that might possibly figure in a Ramsey-generalization it must hold true that had x in fact been a Prime Minister, x would not exceed seven foot five in height. It may well be true that no actual Prime Minister is in fact currently taller than that, but it would not appear to be something we could rule out. However, if a Ramsey-generalization were a valid law it would follow that had x been Prime Minister and the Ramsey-generalization met, x would have been less than seven foot five in height. Counterfactuality, then, seems to be a sufficient condition for lawfulness.

Laws may exist, the objection runs, without being instantiated. It is conceivable that there are laws that are vacuous because not actualized in any instances. An empty law is not the same as no law. Possible instances, had they been actualized, would have conform to the law. But empty regularities which are laws, are indistinguishable from empty regularities which are not laws.

Other objections as well may be raised against the regularity theory. The criticism of the regularity theory, however, can be summed up in the following three points. Laws explain their instances, including regularities; the latter have no explanatory power. Individual facts count as evidence of the existence of a law but the law is more than the sum of its instances. It is possible for a systematic regularity to diverge from the laws that exist.

Counterfactuality indicates the difference between contingent and nomic regularities.

The maximalist position

In reaction against the obvious deficiencies of the minimalist theory, philosophers have developed another which could be called 'the *maximalist* position'. Common to the various versions of this position is the belief in the holistic properties of law. The law is something other and more than its actual instances: laws are necessary connections, or at least relations, between universals.

In line with this view, some philosophers have held that law should be conceived as based on a concept of logical necessity: it is thus logically necessary that all Fs are Gs. They propose that a law be defined as follows:

Definition: It is a law that F is G iff it is logically necessary that all Fs are Gs.

This proposal is less than persuasive. For what holds with logical necessity, is true in all possible worlds. A possible world is a world which is internally consistent but possibly different from the actual world. Accordingly, no law of nature can hold in every possible world. Included among the possible worlds are all those in which the laws of nature are violated. There are, consequently, possible worlds in which iron never rusts, in which potassium cyanide is not deadly, and in which gold does not dissolve in *aqua regia*. Most causal laws are only true if the nomologically relevant conditions are realized in the world in question. No contradiction is involved in conceiving them not to be. In contrast to this it is true in every possible world that bachelors are unmarried. Given our definition of the term, it holds true in all worlds.

Others, such as Karl Popper and W.C. Kneale, have countered that laws of nature are, of course, only necessary in every physically possible world. The difference between the actual world and other physically possible worlds is simply a difference in boundary and initial conditions:

Definition: It is a law that F is G iff it is physically necessary that all Fs are Gs.

This definition disguises a difficulty that cannot be bypassed. Physical necessity is characterized by physically possible worlds, that are in turn specified as the worlds that have the same laws of nature as the actual world. But how are we to account for the concept of the law of nature by appeal to that very concept? Physically possible worlds are defined precisely as those worlds in which the laws of nature hold. Popper and Kneale's proposal cannot escape this circularity.

A third proposal, which counts among its adherents names such as F. Dretske, Michael Tooley and David M. Armstrong, views a law of nature as a relation between universals. A law that says that all Fs are Gs is not a law specifying a regularity between particular items but a relation between properties of the *universals* F-hood and G-hood. But the relation between properties will itself be the bearer of properties that it shares with relations of the same type. That relation must thus itself be a universal. So to avoid potential difficulties, the theory discriminates between first-order universals, properties of or relations between particular things, and second-order universals, which are properties of, or relations between, first-order universals.

The relation between the property of being magnesium and the property of combustibility is a law. It is a relation between first-order properties. But what is the property of this relation that it shares with other relations which also obtain between first-order properties? That property is nomic necessity:

Definition: It is a law that all Fs are Gs iff F-hood necessitates G-hood.

Nomic necessity is, then, a second-order property. We shall here take nomic necessity to mean that F-hood necessitates G-hood if (1) everything that is F is also G, but where (2) the converse implication does not hold, *i.e.*, that everything that is G is also F. Anything that has the property G-hood will only contingently have the property F-hood. Since necessitation is a relation it is also a universal. And since necessitation is a relation between universals it is a second-order universal. Lastly, since necessitation is a universal, it has possible instances. Its instances are those cases where a's being G is contingent on a's being F.

At first sight this proposal seems to have more merit than its two predecessors. But I find it equally lacking. The analysis fits with the laws of co-existence and those of succession but not with the laws of prohibition. The latter cannot be characterized by any relation unless one is willing to accept the existence of negative universals. Examples include the fact that no particle movement can exceed the speed of light and the fact that perpetuum mobiles of a second order do not exist. Moreover, it is doubtful that it would help the position to claim that such laws should be analysed as a relation between universals where one of the relata is a negative property. This would open up for an infinite number of laws. What no object of a certain type can do establishes itself as a negative property of that kind of object. Hence, it would be a law that, say, electrons do not attract one another, do not attack each other, do not eat, smile, or dance etc.

There are also difficulties with the relation of necessity. The proposal assumes that when we speak of a law it is always a case of a *de re* relation, a necessity between the things themselves and not a *de dicto* relation, a necessity between the propositions that make reference to the relevant items. However, there are good arguments that show that 'law' is not a homogeneous category and that the relation to which the law refers is not the same in every case.

But we have not yet addressed the most serious difficulty. The point of departure for the maximalist position is that laws constitute something over and above the properties attaching to their instances. We can accordingly conceive of empty laws. An empty law is a relation between universals that have no instances falling under them. This standpoint is thus committed to the existence of universals independent of any individual instances. So universals cannot exist as concrete entities in time and space. They exist outside of time and space as abstract entities, ideas or concepts. The maximalist position not only presupposes scientific realism but also entails radical Platonism. An ontology that has to explain a law of nature by reference to the existence of abstract concepts has unwanted corollaries. Thought has pulled itself up by its own bootstraps to become the true reality unless, of course, one insists, like Armstrong, that uninstantiated universals and therefore uninstantiated laws do not exist.

The constructivist position

But perhaps our spotlight has been misdirected. Up till now we have been approaching the issue of laws of nature as though it constituted an ontological problem, assuming that laws of nature concern *de re* and not *de dicto* modalities. However, certain philosophers have proposed that laws of nature also concern relations between our statements and not simply those between the things themselves. Frank Ramsey thus sought to avoid what we have termed Ramsey-generalizations and David Lewis has since followed suit and offered a systematic account of the concept of law.

From Newton's laws a great number of other laws may be derived. We have all heard of Kepler's laws of planetary motion, Galileo's law of free fall, Hooke's law of elasticity etc. Such laws are less general, not so fundamental, their scope restricted to fewer phenomena. But they can in fact be derived from Newton's three laws of motion, which accordingly are far greater in scope and thus more general. So Ramsey and Lewis conceive of Newton's theory as an axiomatic system in which laws are axioms or theorems. Regarding the lawful relation between two properties F and G it may be said:

Definition: F and G are connected by a law iff: (i) all Fs are Gs, and (ii) sentence (i) is an *axiom* or a *theorem* in an axiomatic system that encompasses the entire history of the universe and is the maximal combination of strength and simplicity.

By 'strength' Lewis means that all the propositions have determinate informative content and that a proposition such as 'All mammals give birth to live offspring' has greater strength than 'All mammals in Africa give birth to live offspring'. The explication of 'simplicity' is more complicated. There are a number of ways in which a set of data may be described axiomatically. Which qualifies as the most simple? Should simplicity be measured in terms of the number of primitive expressions in each axiom or according to the number of axioms? One way of resolving the issue might be to say that a system with fewer axioms is simpler than one with many; but it might just as well be said that a system with more axioms and fewer theorems will be simpler than the converse case. There is no unequivocal answer. Who is to say that strength and simplicity combine in a uniform manner across different axiomatic systems? The antique theory of the four basic elements, fire, earth, water and air was much simpler than modern atomic theory, while the converse surely applies with respect to strength. Different values for strength and simplicity open up for many different maximal combinations.

Laws of nature have to do with the relations between properties, but physical regularity is not enough for a law. If we look at the definition it tells us that:

(1) If *F* and *G* are lawfully connected, then given (i), all *F*s are also *G*s,

but

(2) Because of (ii) the converse implication does not hold: If all *F*s are *G*s then *F* and *G* are lawfully conjoined.

What is lacking has nothing to do with the things themselves but with the way in which we systematize and logically connect our sentences. A proposition does not express a law because it refers to a law as a real universal relation but because it has a place in an axiomatic system. A law of nature is in part determined by the way in which we construe the world.

The constructivist is right in maintaining that it is, in principle, possible to formulate an infinite number of hypotheses on the basis of a specified quantity of facts, and it would be wonderfully convenient were we able to determine the correct formulation by reference to certain formal criteria. But our hypotheses are empirically underdetermined and one adequate formulation is no less plausible than any other. However, uniform formal criteria for the identification of laws of nature do not appear to be available. Furthermore, the constructivist seems to be faced with the predicament arising from the circumstance that much scientific knowledge is not expressible in an axiomatic system – as, for example, the fact that potassium cyanide is deadly. Nor is it easy to understand why the law that all gold is dissoluble in *aqua regia* needs to be expressible in an axiomatic system to be a law of nature. It would seem, rather, to be a reasonable constraint on a law of nature that it be what it is regardless of whether it is expressible in an axiomatic system.

The deconstructivist position

We seem to be left with no satisfactory account of laws. Is there really no avenue as yet unexplored? I would suggest that there is, for I shall now go on to propose a quite different position, which I shall call the *deconstructivist* theory. Its core claim is that not all laws are laws of nature. Fundamental laws such as Newton's three laws of motion do not really exist. Newton's laws have to be deconstructed if they are to be capable of explaining anything at all and of describing concrete causal connections. Scientists are invariably confronted with the task of explaining individual phenomena occurring in nature, and need therefore to take account of factors impacting on the presence of such phenomena. Among a given species of phenomena many concrete circumstances will vary from one instance to another. In other words, the behaviour of a given phenomenon is dependent on its connectedness to other phenomena. Fundamental statements of law are limited by *ceteris paribus* clauses which have to be adapted to a particular model before they can be used to describe individual phenomena.

We need to distinguish between *fundamental* laws and *phenomenological* laws. Nancy Cartwright has drawn attention to the fact that Newton's laws describe with accuracy only a physical system containing two bodies if the system is not subjected to other forces. In fact, laws have countless exceptions when other of the forces recognized by physics figure in the system, and once the system contains three bodies the laws no longer hold. In having a range of exceptions they resemble other fundamental laws. Previously, philosophers of science distinguished between theoretical laws and phenomenological laws. That dichotomy turned on a dubious epistemological distinction between laws governing what is observable and laws governing what is not. But the fundamental laws Cartwright has in mind are not that kind of theoretical law. For the distinction between phenomenological and fundamental laws in physics is a distinction between descriptive and explanatory laws. For example, Balmer was able to give a mathematical description of the hydrogen spectrum. He formulated a phenomenological law but it offered no explanation of why hydrogen emits light as it does. That was delivered only by Bohr's ideal model for the hydrogen atom, on which the atom consists of a positive nucleus and a negative electron which travels in stable orbits round the nucleus. When the electron jumps from an outer to an inner path the atom emits light with a frequency corresponding to the energy differential between the two trajectories. Bohr's fundamental law for the quantum leap explains the Balmer series, but the law is not a true one if conceived of as explaining the real hydrogen atom. It is only true of Bohr's model of atomic structure. Balmer's phenomenological law gives, by contrast, an exact description of hydrogen's emission of light.

Fundamental laws refer only to theoretical models; they never give a true description of observable phenomena. As against this, phenomenological laws describe the world correctly but perform no explanatory function. Cartwright's position can be summed up in four main theses:

1. The remarkable explanatory power of fundamental laws is no argument for their truth.
2. The way in which these laws are used in explanations speaks rather for their falsity. We explain by means of *ceteris paribus* laws, by the composition of causes, and by approximations that improve on what is established by fundamental laws.
3. The semblance of truth derives from an inadequate explanatory model on which fundamental laws are directly linked to reality.
4. The traditional position has to be replaced by a simulacrum model.

Cartwright argues that fundamental laws are false, and that if they are not to be false, they must be rendered serviceable *qua ceteris paribus* laws. Once so relegated, they no longer enjoy the status of laws proper since they only apply under special and idealized conditions.

But how are we precisely to understand *ceteris paribus* clauses? Consider, as an example of a fundamental law, Newton's law of gravity. According to Cartwright it is false because it rests on the erroneous assumption that the mass of all bodies are in every case concentrated at one point. But no mass is so concentrated in reality. So in order to preserve the law of gravity from falsity it is necessary to prescind from actual circumstances and formulate the law with the rider that it only applies to bodies without extension. Such bodies do not exist in the real world, only in abstract models.

Cartwright's argument represents an important contribution to the understanding of the concept of law but I disagree with her analysis on two central points. For I shall argue that some fundamental laws do not express facts, and therefore do not contain *ceteris paribus* clauses, whereas other fundamental laws as well as phenomenological laws state facts, and in virtue of so doing always contain a *ceteris paribus* operator.

There is no reason to say that *ceteris paribus* laws are, as such, not real laws. Cartwright is right insofar as we may conceive of them as fundamental laws with restrictions on them. These refer only to models. But we must distinguish between *ceteris paribus* laws that are fundamental laws with restrictions, and those that are phenomenological laws with restrictions. The latter meet a constraint on any concept of law according to which universality without *ceteris paribus* clauses is not a necessary or essential property of laws. The difference between such fundamental laws and phenomenological laws rests not in a difference in the restrictions imposed upon them, but in fundamental laws being used to explain their phenomenological consequences. Newton's law of gravity explains Kepler's laws, which describe the actual orbits of the sun by the planets. But Kepler's laws are simply generalized descriptions that are true under certain typical conditions to which the actual conditions of real individual planets only approximate.

Science begins by describing what typifies the individual connections between disparate phenomena. There may be both causal and structural connections. Should the scientist succeed in identifying a constant connection, he will have discovered a phenomenological law. Next, science can tell causal stories about individual phenomena by reference to specific *ceteris paribus* laws that obtain in typical conditions. But it is also clear that such *ceteris paribus* clauses may be so numerous that the so-called law describes only one phenomenon. Lastly, the scientist looks for a general explanation of the causal or structural connection that he has found to be descriptive of the individual phenomena. With such an explanation at hand, he has framed a fundamental law.

Phenomenological laws are not Ramsey-generalizations. Whereas Ramsey-generalizations were formulated as empirical *ceteris paribus* statements, the phenomenological *ceteris paribus* laws also include laws of causation that can be expressed with the aid of generalizing causal statements. Both types of statement are true under certain rider-enjoining circumstances, and given that qualification, the statements in question are universal in scope. They apply to all the instances that satisfy the rider. And even if science seeks to identify causal laws applying to natural kinds, it is not exclusively *qua* phenomenal type that one phenomenon causes another: it is through its figuring as an exemplar of a specific kind in conjunction with certain nomologically relevant conditions. An instance of a natural kind only ever functions as a cause if certain circumstances obtain. It is these circumstances that make the instance in question a cause, and if they are sufficiently alike from the one occasion to the other on which we encounter instances of the same kind we will have nomologically relevant conditions and therefore a law of nature.

Examples are legion. The heat from a hotplate does not always cause water to boil at 100°C. That occurs only if the pressure is one atmosphere and the water is fresh and free of impurities. Nor must the water be heavy water in which hydrogen does not occur in its usual isotope. A match does not ignite simply because it is struck against the sulphur on a matchbox: the match has to be dry, the friction sufficiently substantial, and oxygen present in its vicinity. Such is the case with all causes: their functioning as causes is contingent on the fulfilment of a vast number of conditions where such conditions are *nomologically relevant* in the sense that only when they are fulfilled the cause, *qua* particular event, will always produce another particular event as its effect.

In many cases we can directly observe individual causes. A woman on a pedestrian crossing is hit by a car and collapses onto the road. Lightning strikes a tree and splits its trunk. Causation is a basic ontological category, and there is no reason to believe that it can be fully understood in terms of other concepts such as extensional regularities, counterfactuals or probabilities obtaining between different phenomena: their exemplification in a correlation between two types of phenomenon may, however, be taken as evidence of a causal connection. What we do believe is that the lightning in the particular case was a causally necessary determinant of the tree's splitting, that its being so struck necessitated its destruction, and that the lightning strike was causally prior. No other candidate concept can capture these properties of individual causal connection. The notable difference between Ramsey-generalizations and causal facts is thus the counterfactual aspect. The implication that follows from every individual causal statement is, that had the cause not occurred, neither would the effect. It is this counterfactual relation that shows us that the triggering event was causally necessary. The converse counterfactual expresses necessitation, and so causal priority cannot be expressed as a counterfactual relation.

We can explain an individual event by reference to its individual cause – we cannot explain an individual event by reference to a causal law. The causal law is simply a generalized description of individual causal connections, and we can only explain a fact by reference to an independent fact. By contrast, one law can explain another law.

A causal law is thus a generalization over a particular class of causal facts under nomologically relevant conditions. Let us express the causal law as C causes E where C and E are types of property of particular phenomena, events, etc., x and z:

Definition: '*Ceteris paribus*, C causes E' expresses a causal law of nature iff $(x)(y)(z)(C(x)\ \&\ K_j(y)$ causes $E(z))$, where y runs over all nomologically relevant circumstances K_j.

Since the relation 'is the cause of' cannot be represented by any other known symbol, this nexus has consequently its own distinctive logic, which is not identical with ordinary predicate logic. We also observe that the cause always functions in conjunction with certain nomologically relevant conditions. When we distinguish between them, it is solely on account of our epistemological interests. If a car crashes on a bend it might be because the motorist was driving too fast, because the car hit a pothole and because the steering wheel broke. A combination of factors caused the accident. Which of them we identify as the cause will often depend on whether we represent the car manufacturer, the highway authority, the police or the relatives of any injured party. I think typically the point here would be that we atomize the holist web of facts that constitute the state of the universe at any instant for basically pragmatic purposes, and when we do so we select certain particular ones as especially causally pertinent to others. We speak of these facts as those which caused that fact, but we recognize silently (*i.e.* pragmatically) in the context a million *ceteris paribus* clauses with respect to other facts which are taken for granted.

The other problem with Cartwright's analysis is her characterization of fundamental laws as false. She concedes that it is unintelligible how such laws come to be false while possessing considerable explanatory power. And she is not alone in being mystified. For if they are capable of being false, they must, in principle, also be capable of being true. How can it, then, be reasonable to say that they are *always* false? And how can she deny that fundamental laws have descriptive content when they are false? If they are false, it is because something actually makes them so.

I contend, contrary to Cartwright, that some fundamental laws, such as Newton's three laws of motion, are neither true nor false because they do not represent anything in the physical world. Instead they are essentially *rules*, explicit *linguistic rules*. Others may be said to be true with respect to an abstract and idealized model. Cartwright's analysis is correct inasmuch as it is true that no fundamental laws can be applied directly to the description of concrete physical systems. Every nomic explanation must always take account of the specific circumstances governing each individual case, which are unique from one place and one exemplar to another. So we construct a model of the system in question to gain an insight into what makes the fundamental laws amenable to formulation in terms of a causal description of the concrete phenomenon. For a model adjusted to the actual circumstances is an ideal representation of a concrete situation. This is made possible by the use of what Thomas Kuhn calls 'exemplars' – and which might profitably be called *standard models*. They prescribe how the application of fundamental laws to concrete systems should proceed on the basis of the scientist having come to regard them as idealized descriptions which, allowing for certain approximations, may be applied to a specific situation.

We can take classical mechanics as an example. Ordinarily Newton's three equations of motion are used to describe a mechanical system. Characteristic for the mathematical formulation of these equations is the use of the term 'force'. But there exist alternative formulations of the term such as those, say, offered by Joseph Louis Lagrange and William Rowan Hamilton. The notion of a force is, in their equations, replaced by one of energy. If the content of Newton's three law statements had been true, one would have expected the notion of force to be indispensable to any description of a mechanical system. Since, then, its role may be filled by quite other concepts, Newton's equations of motion cannot refer to laws of nature but must be regarded as an explicit definition of both the notion of force and of that of natural motion (the law of inertia).

A mechanical system can take many forms and be subjected to many different conditions. In consequence, Newton's laws of motion cannot be used to describe an individual phenomenon. Nevertheless, the description of any particular phenomenon will correspond to one of the various basic forms represented by a standard model. Whenever Newton's laws are applied in the recounting of a causal story, they are so applied by dint of a model. The student becomes acquainted with Newton's laws by applying the equations to standard models such as, for instance, uniform circular motion, free fall, the inclined plane, the pendulum, coupled pendulums, harmonic oscillation, etc. Newton's laws of motion furnish the researcher with a language and a set of linguistic rules which she can apply to a model in telling a causal story, and the standard models each impose different *ceteris paribus* clauses on the actual use of such rubrics. With the introduction of these riders, we have (or obtain) laws that apply to theoretical models. *Ceteris paribus* laws and models go hand in hand. Precisely how is the topic of the next chapter.

But for the moment, to illustrate the present point, let us consider the workings of a grandfather clock. Newton's laws of motion are theoretical laws; they provide us with a language, whereas his fundamental law of gravity expresses the fundamental forces to be conceived of as operative in an idealized model of this type of system. The real motion of the clock's pendulum, however, conforms to the causal laws governing pendulums, but it also exemplifies quite specific degrees of friction, gravity and equilibrium; similarly, the clock has a particular location which makes the pendulum's oscillations different from those of pendulums in other clocks of that kind. The pendulum is in one sense like any other, and yet is different from them. The particular milieu of the pendulum must be taken into account if the model is to be used to describe this precise pendulum. All relevant and concrete conditions must be built into the model if the story told is to be accurate. The law of gravity must be deconstructed and adapted to a more specific model of the system before the causal story can take off, *i.e.* the fundamental law has to be broken down, its *ceteris paribus* conditions dissolved and concretized within a model before we are in a position to explain the movement of the pendulum sited at a given location.

Newton's laws of motion can be compared to the rules of a natural language. No such rules are true or false. They are primarily prescriptive and may be valid or invalid. They determine what counts as correct or incorrect usage. It is concrete descriptions of things that are that are true or false; individual sentences are true or false when used to make determinate statements about the world. So it is too in the case of Newton's laws of motion. A particular explanation relates to an actual system and may accordingly be true or false. The laws, by contrast, furnish us with the rules for linguistic means of representing the system, and are in consequence neither true nor false. The movement from phenomenon to law requires rules of entry, and that from law to phenomenon, rules of exit. The former movement we might call interpretation, the latter, explanation, and the fashioning of the rules is in part dependent on prevailing observational and experimental practice.

Social laws

In the social sciences two basic sets of epistemic interests collide. Should society and its institutions be understood on the basis of causal laws as in natural science, or on the basis of the human use of signs and symbols? The difference in perspective offered by each is dependent on the *ontological* conception of society to which the researcher is wedded. If she configures society as a set of highly complex, dynamic elements in a state of ongoing mutual interaction, she will seek to explain social phenomena by formulating social and economic laws that hold of society; if, instead, she holds that society is most appropriately understood as governed by juridical and cultural rules and norms, she will make the intentional and semantic aspects of society pivotal.

Each of these ontological perspectives may be further partitioned into two positions that cut across categorizations of lawful and semantic. One position is called *holism*. On this view society is conceived as being more than the sum of particular individuals and their interrelations, and an explanation of individual actions is afforded by reference to more general principles. One possibility, then, would be to characterize the system by reference to certain quite general economic or institutional properties that determine individual agency. Another possibility is to ascribe to the system certain functional needs, intentions or aims and to claim that a particular phenomenon (the behaviour of the parts) occurs because it meets these needs, intentions or aims.

The alternative position is called *individualism*. Here one assumption is that every event in society can be understood through knowledge of the some causal properties of each individual, so that a person's social behaviour can be grounded in psychological dispositions or other features of personality. Another assumption is that social phenomena must ultimately be understood, not in terms of explanatory features of causation, but as meaningful and intentional actions governed by free will.

These positions are often referred to as methodological holism and methodological individualism, respectively. But is this really an apt way of identifying metaphysical disagreement, considering that the disagreement turns on disparate ontologies? Choice of perspective reflects certain epistemic values and goals, but the choice in the present instance is mainly between different ontological conceptions. It will be argued here that holism and individualism need not constitute metaphysical alternatives and that both are scientifically respectable so long as they satisfy the constraints common to all scientific research.

Karl Marx's doctrine of the impact of economic laws on social relations is a holistic theory. According to Marx the means of production creates social consciousness. It is not, he contends, human consciousness that determines human life but social existence, rather, that determines consciousness. People enter into the relations of production without having given their consent, and it is the resultant relations in their entirety that constitute the social structure of society. Material production is thus the driving force in historical development in that the forces of production, *i.e.* equipment and machinery along with human technical knowledge, come into conflict with the conditions of production. Economic laws cannot, then, be reduced to psychological laws (indeed, the other way around), and social regularities are more than just individual social facts.

The objection levelled against the holist by the individualist is that there are no mechanisms that cannot be modified by individuals who seek to do so, once cognizant of their existence. Social mechanisms cannot, then, be independent of individuals. Further, she will wonder how such laws can exist. For holists, social laws will obtain even in the absence of societies or people. It must be possible for them to be empty. But then, where are they located? the individualist will ask. But this objection against holistic laws in the social sciences becomes inapplicable if, instead of giving them an ontological interpretation, we identify Marx's laws with linguistic rules – as we did with Newton's laws of motion. None of them refers to actual regularities. Instead, these linguistic rules furnish us with a language that we can use to describe conditions obtaining in society. Theories such as Marx's equip us with a conceptual scheme that categorizes and defines our realm of objects.

An individualist theory sees human subjects as rational social agents whose social behaviour can be described by reference to individual mental states. Society consists of individuals, and social phenomena are explained by appeal to a conception of the attributes that such individuals possess. The theory of rational choice ascribes to humans a capacity for rational behaviour, and so action is explained by setting up an idealized model for the behaviour of rational agents. Such a model consists in ascribing to the agent (i) a list of ordered preferences, (ii) full information about the actions available to him or her, and (iii) the facility of a perfect inner calculator. These properties place the agent in the position of being able to make rational choices from among various alternatives. A rational action is one that will maximize predicted utility value. The theory thus conceives of social agency as the outcome of a mechanical calculus.

A rational choice is made from among a system of ranked preferences whose realizability is calculable. Preferences and choices of action naturally vary from one individual to another. In other words, the individualist can frame only a model that is adjusted to the particular individual, and with actual agents describable only by reference to *ceteris paribus* laws. However, the individualist is without any general explanation of the choice of values and preferences that the holist makes part and parcel of his theories.

The difference between holistic and individualist perspectives is effectually a reflection of particular idealizations and abstractions that need not be the expression of a deeper ontological divergence. The distinction between holism and individualism is no different from the parallel distinction between fundamental and phenomenological regularities in natural science; both perspectives view human agency as governed by economics, or by psychological laws arising from the causal properties attaching to social phenomena. Marx's theory and the theory of rational choice may be seen as examples in social science of fundamental laws and phenomenological laws, respectively. The dispute between holists and individualists rests on the misconception that both fundamental laws and *ceteris paribus* laws are objective social laws, and that statements corresponding to them are true or false in virtue of their reference to these laws. If, instead, we come to understand some fundamental law statements as linguistic rules the disagreement will evaporate.

Social rules and conventions

There are other social scientists who judge the theorist whose explanations appeal to social laws to have completely missed the fact that actions are usually performed with a particular end in view. Human actions are not so much causal as intentional. They are directed towards an end. To capture the underlying intention the researcher needs instead to grasp human actions as rule-bound, *i.e.* as governed by a predetermined pattern that aligns the actions with others that share the same pattern. Many of these rules underpin our practical experience in the sense that our choice of a language for describing intentional behaviour commits us to rules which shape the ways in which we apprehend the world. If I know the rules of football I recognize that a player has scored a goal if the ball lands in the net without the player being offside. And if I am familiar with the rules for buying a car I know that the vehicle of my choice is mine if I pay for it cash down or sign a hire purchase agreement.

But in what way does a prescriptive rule differ from a descriptive law? We can define a rule as follows:

> A *rule* is a norm, a uniform and generally accepted prescription, which states what action is appropriate in some more nearly specified type of situation.

It has to be uniform in the sense of being consistent and occasioning the same routine each time we follow it; it must be valid over a period of time, not valid the one moment and invalid the next; and it must be the same for all the specific situations it covers. It must also be systematic in the sense that it does not conflict with other rules, and a prescription that is not followed by persons in the situation for which it is specified ceases to be a norm.

Normally we distinguish between several species of rule. *Constitutive* rules are such as define what it is to perform a particular act. The identifiable action is not performed if the rule is not followed. I fail to buy a car if I simply get into it and drive it out of the showroom without paying or signing a hire purchase agreement. I steal it. *Regulative* rules, by contrast, determine what counts as legitimate action. If I choose to pay for my car by instalments the terms for buying on credit are down in writing in the contract along with the consequences of being in default. I can choose to comply with the terms and so act in accordance with the law. But even should I be in breach of contract I retain title. The dealer cannot repossess the car without legal authority to do so.

Our understanding and appraisal of human conduct rests substantively on institutional rules and established law, and it is in virtue of such rules and laws that we occupy social roles such as that of general, bishop, employer, employee, taxpayer, road-user, and football coach. Social rules do not merely define our actions but associate those actions with particular roles. The performance of certain actions is linked by definition to the occupancy of certain roles. In many cases the understanding of the action is bound up with the corresponding role and following the rule defines the role of the actor. Even though it would be to our advantage to break certain rules, we regard one another's conduct as rational to the extent to which it is in conformity with accepted rules. All other things being equal, it is most rational to comply with the rules of the road in order not to come to harm in the traffic.

The intentional perspective also has its individualist variant. This perspective views human social behaviour as comprising willed acts, with the individual's particular characteristics and mental constitution figuring in his or her conduct. Even if we occupy a variety of roles, we fulfil them differently. A general does not only play the role of general; she may also play that of mother, wife, friend and colleague – and perhaps is involved in the social work of the parish. She may well have particular tactical and organizational abilities which enable her to put her personality and moral standards into play in her relations with subordinates; this will make her decisions and orders different from those of another general. Her actions in a state of war would thus be coloured by her knowledge, abilities, values and not least her autonomous choices.

The relation between social roles and personal choice displays a pattern matching that of the relation between theoretical laws and causal laws. Social rules are conventions. In contrast to causal laws, rules might have been other than they are – for they are our creation. The individual may not be able to change them, but collectively we can. Society's rules and laws are instituted and dissolved whereas the laws of nature are constant. Perhaps the laws of nature do not last forever since they might conceivably undergo change across aeons of time. We do not know. But irrespective of such limitations that are due to a choice in how we describe behaviour, theoretical laws are, in line with juridical and social rules, prescriptive definitions. We can make reference to common social rules in an explanation provided their scope is curtailed by the bringing into play of the individual's values, motives, desires and choices. The introduction of such values, attitudes to life and personal choices into intentional experience means the analogous imposition of certain *ceteris paribus* clauses on social rules.

We have now looked at two fundamental perspectives that rest on the respective causal and intentional understandings of the materials of social sciences. These perspectives represent two distinct ontological positions. The one cannot be said to be more correct or 'natural' than the other. They are complementary descriptions where one may be more rationally warranted in a given situation than the other. So long as the requirements of scientific research are met, each will be able to contribute to the overall understanding of life in society. What *is* important are the two constraints that each position must meet: (a) the hypotheses that purport to explain concrete phenomena must be justified using the methods that science makes available, and (b) the hypotheses must be so framed that they are both general in form and can be used in the endeavour to understand a particular situation.

Within certain sectors of the human sciences that have to do with history and artistic production the latter requirement is regarded as in capable of being fulfilled and without scholarly interest. Idiographic interpretation is principally characterized by the fact that it bypasses the general and abstract and seeks to interpret instead the particular and individual.

Idiographic interpretation

It is the task of natural science to provide individual explanations in terms of general concepts whose meaning is determined by a comprehensive set of linguistic rules. However, it is also reasonable to say that research in the humanities also seeks the typical in the individual, rendering that activity scientific in character and so of general import. It would scarcely be possible for us to apprehend a thing merely in terms of what is utterly distinctive about it. Plainly, every property evinced by a person or an entity can in principle be shared by other persons or entities: all predicates and relations can be instantiated by more than one person or thing. The exact combination of properties will vary from one person to another, and from one item to another, but that has no systematic import. Every description must be couched in a language whose meaningfulness is publicly accessible because the conditions for its use are publicly available. The same applies in the case of figurative or metaphorical uses of language, and it is the role of the human sciences to uncover these publicly accessible conditions. Staying with the example of hermeneutics, we might say that a psychoanalytic interpretation of a novel will seek, via an analysis, to attain an understanding of the characters that will bring out the typical and general themes of the book, given that the individual conditions for a particular metaphorical reading are met. Psychoanalysis is based on the general principle that particular psychological patterns inform the development of every individual, and manifest themselves as shared symbolic representations figuring in dreams and other unconscious behaviour. Were it not for the assumption that there are such shared generic features, psychoanalytic interpretation would be devoid of scientific interest.

In an appeal to the laws of nature we usually have in mind universal causal laws. But as we have seen, these are really *ceteris paribus* laws. Espousing this position makes a person an individualist where the laws of nature are concerned. For the individualist, causal laws are merely empirical generalizations based on causal facts attaching to a particular type of individual process. The individualist holds that the holist's concept of intrinsic laws of nature is an unfortunate hangover from the time when scientists believed that God had created all that exists, including the laws of nature that enjoined things to behave in a certain way. A law of nature is not some holistic entity which individual processes express, participate in, or directly manifest. The individualist denies that there are laws of nature taking the form of universal structures that, from the hand of God or nature, determine the way individual processes are to go. He asserts instead that every law of nature is simply a generalization of the conditions governing a given type of particular causal process. Given that conception of law, there would seem to be far less that separates natural science from the human sciences. Such systematic generalizations may be applied equally appositely to both intentional and interpretive explanations.

In the same manner as he does with respect to natural science, the individualist will be able to speak of *ceteris paribus* laws in the human sciences; not necessarily in the form of causal laws but in the form of intentional, interpretive or structural rules which figure as generalizations of acts or meaningful features of particular objects. A *ceteris paribus* law may, as we have seen, be characterized as 1) a relation between two types of relata which may be specified independently of each other; and 2) a quantification over the domain of individual entities the relata stand for, on condition that a *ceteris paribus* proviso is satisfied. Further, the relation does not necessarily specify a causal connection but may represent various kinds of connection depending on whether the quantification governs a class of natural kinds or a domain of intentional or intensional entities. Such a law is indifferent to whether the particular domain of entities that is quantified over under the restriction of a *ceteris paribus* clause is given by a causal, or some form of representational, relation. The relation may well be representational and, if it is, the boundary separating nomothetic natural science and idiographic human sciences dissolves. The former may very well be idiographic and the latter nomothetic.

For instance, the symbolic meaning of the angel and the dove appearing on a beam of light that traverses glass or a window to strike the head or breast of the Virgin Mary constitutes an iconographic law in medieval paintings that links such icons to the Annunciation. We might here follow Erwin Panofsky and, *mutatis mutandis*, refer to typology and the history of types. Typology is the study of the principles that underlie the choice and presentation of motifs as ideals, pictorial narratives, symbols and allegories in which essentially human thoughts and feelings find expression in particular themes and notions under varying historical circumstances. As distinct from that, the history of types examines the ways in which these specific themes and notions have come to expression in objects and events in differing historical conditions. Mediaeval depictions of the Annunciation illustrate the general convention of using light and the dove to symbolize conception and the Holy Spirit.

What differences there may be between the natural sciences and the human sciences in their respective emphases on the universal and individual should be understood less in terms of their ontological premises than in terms of their epistemic interests. For it is because the sun is a sine *qua* non for the preservation of life on earth that astrophysicists are particularly interested in describing solar physics in detail. The optimum would be to give a precise idiographic description of the sun's evolution by including all those features that are required for the elaboration of a minutely detailed mapping of the sun's history. But we have no comparable interest in gaining a similarly detailed understanding of the other stars. A model of the sun is, accordingly, so much more complex because it involves the sun's individual properties: these include its exact age, shape, size, mass, mass distribution, rate of rotation, magnetic field, sun spots, composition of elements, etc. In what concerns other stars we are content just to know their more general physical properties and evolutionary laws.

In the human sciences such as history, literary studies and art there is undoubtedly a tendency to focus on the individual features of persons, characters and artworks. We are interested in understanding what is peculiar to them, something that is in part determined by specific factors, *viz.* the social, historical or literary context into which the characters and works themselves fit. Bad art represents characters as types rather than as individual personalities. The education of the literary critic thus points him towards seeing the characters in the round and in all their variety. But even in the most hidebound idiographic sciences there is at the same time a thrust towards the identification of the shared and general. When such considerations are in focus, the individual is regarded merely as an exemplification of the general. So it is, then, that the archaeologist sorts his finds into those from the stone, bronze and iron ages; the historian strives to render a general account of the condition of the peasants after the abolition of adscription, and of the scattering of farms prior to the introduction of crop rotation; likewise the literary historian examines, say, the metric structure of broadsheet ballads.

The idiographic description that aims at being scientific is pivotal to the distinction between accidental and essential properties. The individual features picked out must be essential and not accidental. Unless they are, their inclusion in the account fails to advance scholarship in what concerns the understanding of the work. Far from all the properties that attach to an individual are relevant to an idiographic description. Without such a constraint, we would be able to cite the list of names in the telephone directory and call our list science. The qualities of a character in a novel are important if they are decisive for its narrative structure, sequence of progression or plot. If the author portrays a given character as bespectacled, it will, for most readings, be of no interest to a scholarly interpretation but it might in a particular context. Idiographic analyses are more selective in scope than an inventory of the totality of properties figuring in the work – a distinction must be drawn between relevant and irrelevant features.

A given property is only significant given a particular perspective. Every theoretical understanding builds on abstraction and idealization, irrespective of whether its object is nature or human life, and an interpretation is only of scientific interest if it invokes general concepts, or refers to general theory. The hallmark of a scientific interpretation is its systematic character. This feature is what marks off a theoretical grasp from a personal reading. I am no more a literary scholar when I recount the storyline of a book to my daughter, than I am a physicist when I tell her that snow is frozen water. Explanations can be true without having scientific value. Individual understanding is not identical to idiographic understanding. The former attaches to the person doing the understanding. The latter draws its character from the properties of what it is that is understood: understanding is arrived at through the deployment of general and abstract concepts. In sum, the boundaries between idiographic and nomothetic description are fluid and not fitted to mark off two distinct approaches to the world.

References

Armstrong, D.M (1983), *What is a Law of Nature?*, Cambridge University Press, Cambridge.
Cartwright, N. (1983), *How the Laws of Physics Lie*, Claredon Press, Oxford.
Hollis, M. (1994), *The Philosophy of Social Science*, Cambridge University Press, Cambridge 1994.
Kneale, C.W. (1950), 'Natural Laws and Contrary-to-fact Conditionals', *Analysis 10*, reprinted in T.L. Beauchamp (ed.), *Philosophical Problems of Causation*, Dickenson: Belmont, Cal. 1974.
Lewis, D. (1973), *Counterfactuals*, Basil Blackwell, Oxford 1973.
Panofsky, E. (1982), 'Iconography and Iconology', in *Meaning in the Visual Arts*, Chicago University Press, Chicago, Ill.
Popper, K.R. (1959), *Logic of Scientific Discovery*, Hutchinson, London.
Ramsey, F.P. (1978), *Foundation*. Ed. by D. M. Mellor, Routledge & Kegan Paul, London.
Salmon, W.C. (1984), *Scientific Explanation and the Causal Structure of the World*, Princeton University Press, Princeton New Jersey.

8 Theories and Models

Can we say of scientific theories that they are true or false? A commonly held conception has it that while the starting assumption of a scientific realist is that theories are true or false, an instrumentalist or scientific antirealist denies that theories have an epistemic content that renders them true or false. The antirealist contends, then, that theories are merely heuristic tools in the service of thought, they organize what is otherwise an unsurveyable amount of data, or else epistemic tools for the generation of predictions.

A more sophisticated position has it that scientific theories are neither true nor false while not going so far as to say that theories are merely instruments of prediction. Scientific theories describe not the world, but abstract objects in a model. I shall be arguing that theories are, essentially, sets of explicit linguistic rules for the construal of various terms – terms that refer to properties of the objects figuring in the model. A scientific theory defines, then, the meaning of the relevant terms. Newton's theory builds on the three so-called laws of motion. None of them are literally laws, in the sense of being descriptive statements about the world, but are instead specifications of the meaning of the concepts of force, mass, velocity and acceleration, acquired through scientific practice. In consequence, a scientific theory becomes a way of recounting a causal narrative involving particular entities such as the orbit of mars round the sun, the free fall of a plummet from the leaning tower of Pisa, the pendulum in Jens Olsen's world clock, the tide at London Bridge, etc.

This position is relevant not least to research in the humanities. There too, I would contend that theories furnish us with the linguistic means to recount a story with point and substance about actual persons, artworks etc. Psychoanalysis, structuralism or any other theory within the interpretive sciences is neither true nor false but provides us with a vocabulary in which to talk about a particular subject or a given context – such as, say, Oliver Smith's neuroses or Kafka's *The Castle*.

Abstraction and idealization

Laws of nature govern natural kinds – indeed, some philosophers go so far as to say that a law of nature can only be characterized in relation to the natural kinds of which it holds. If this were so, it would obviously be idle to define a natural kind by reference to a law of nature. But if we can show that natural kinds are represented in science as isolated and idealized objects, resulting from conceptual abstraction, then by the same token, any concept of a law of nature must similarly rest on an abstraction. And if we can show, furthermore, that the theories of natural science give us merely a vocabulary, the ontological interpretation of many so-called laws of nature will crumble.

Reality is an integrated whole – everything is linked up with everything else and marked by relations of interdependence. But as we experience it, the world is split up into discrete units – fragments of a whole. We see objects set in their surroundings, individuals in society and the artwork in its context of meaning. But when we want to describe an object we isolate it from its surroundings, as we do the individual from society or the work from its context. We circumscribe the item, whether it figure in a description or in an experiment, and so sever it from its multiple connections to other items. Whether we are interested in investigating the deflection of electrons in a crystal lattice, the effect of organic solvents on cerebral membranes, or the evolution of blank verse in the Renaissance, we look at concrete items which, in our attending to them or in the course of experimentation, we segregate from their natural milieu. This is perhaps true to a lesser extent of subjects relating to society or the arts, where it is commonly recognized that the properties of people or entities may be identified only in the context of other persons or things. But within these areas too there will always be connections that never come to figure in the relevant description, and which we deem unimportant for the understanding of the phenomenon in the respect we have in focus.

A scientific investigation always focuses on particular properties of concrete things. But each such item can be described in multiple ways. A book is normally defined as an object with printed pages, made of paper and bound in leather or cardboard. But it can also be regarded as a physical object, an item composed of particular chemical combinations, or again as something likely to fetch a high price, as rare, as a gripping read, or as absorbingly interesting. There are sciences which focus on these features individually, but none that addresses them all. We simply disregard all the properties that are not germane to our enquiry. Not only contingent features but sortal properties as well are completely ignored. Any specific item needs to be conceptualized if it is to be the subject of scientific enquiry.

This process involves first *abstraction* whereby the individual item comes to be represented by an abstract object. First a concrete thing is classified as an item of a particular type, then, certain sortal properties are made central while others deemed irrelevant. For not all sortal properties attaching to it are included in a given representation of the item in question. Even the most fundamental physical things such as electrons have a variety of sortal properties: mass, charge, spin, energy, impulse, properties which only rarely or never enter collectively into the understanding of the processes which involve such electrons. Science, then, studies concrete things, processes, phenomena, by contriving their representation as abstract objects evincing just one or very few properties.

Finally the scientific object is a result of *idealization*. We idealize the object: we let an ideal construction represent the concrete individual regarded by abstraction as exemplar. We seek to present a given object as an ideal instance of a natural or nominal kind. Properties that the object does not have *qua* its membership of a particular kind do not enter into the representation. In many cases the actual object will be subject to contingencies that make it difficult – if not impossible – for it to be represented by some ideal object. The researcher thus specifies certain *ceteris paribus* clauses – conditions which the individual instance must satisfy to qualify as an item of a particular type. For instance, in Newtonian mechanics objects are represented as ideal masses without extension and in economics agents are referred to as perfect calculators.

The objects with which science is concerned comprise both concrete and abstract entities, with the latter representing the former. The researcher refers to abstract entities in order to make theoretical claims about the real world. And as we shall see, these entities figure in models which are pivotal to the theories themselves.

Theories

The term 'theory' is used frequently and extensively and has many senses. Here, we are primarily interested in scientific theories. On my view theories are not identical with presumptions or hypotheses. That smoking causes lung cancer is a well-grounded hypothesis but will never be a theory, even were it to be confirmed a million times over. By contrast, Newton's theory has never been a hypothesis. It was a theory from its inception. So prima facie a theory must be understood as a set of principles or ideas that ground a science. The difficulty is to get beyond this first approximation. In fact, there are three more or less elaborated proposals on offer: 1) the *syntactic* interpretation; 2) the *semantic* interpretation; and 3) the *linguistic* interpretation. In my judgement, only the final member of this trio is able to provide us with a persuasive explanation of scientific theories.

The positivists – the logical positivists, note – were the first to give scientific theories a formal cast. The attempt was later dubbed 'the *syntactic* position'. A theory is an axiomatic system, capable of formulation in terms of mathematical logic, and with the axiomatization fulfilling certain conditions:

(i) The theory is formulated in a first-order language L.
(ii) The non-logical terms or constants of L are divided into three different vocabularies: (a) the logical vocabulary comprising logical constants and mathematical terms, (b) the observational vocabulary O comprising all the observational terms, and (c) the theoretical vocabulary T comprising all the non-logical and non-observational terms of the theory.
(iii) The observational terms O refer directly to observable things or observable properties of physical objects.
(iv) There exists a set of theoretical postulates A whose only non-logical terms come from T.
(v) The theoretical terms T are explicitly defined by a correspondence rule C that connects them with the observational terms in O.

These conditions were improved upon over time, with (v) in particular being the subject of serial refinements.

The correspondence rules serve three functions in the syntactic interpretation. First, they define the meaning of the theoretical terms; second, they secure their epistemic relevance; and third, they determine the admissible experimental procedures under which a theory may be applied to the phenomena. If, say, a correspondence rule defines the theoretical term 'mass' as identical with a measure M of an object under circumstances S, and M and S can be described in O, the rule simultaneously ensures that there is an empirical procedure which links the theory to the world, that it is cognitively relevant, and that 'mass' is well defined. However, an obvious weakness of this interpretation is that the meaning of the theoretical terms is reduced to that of the observational terms. This does not sit well with the fact that we tend to see the measurements as a testing of the theory's testable consequences, and that the results attained are interpreted as empirical manifestations of the interaction of the theoretical entities with the measuring apparatus.

There are, today, serious objections to the syntactic position that show this interpretation not to be tenable. The crucial distinction between theoretical and observational terms cannot be sustained: our experiences are significantly theory-laden and theoretical terms are meaningful independent of these experiences. Another problem is that in actual fact relatively few theories can be axiomatized. That can only be achieved if we have a full and thorough knowledge of the interconnections between all the concepts of the theory, and these are unambiguously determinable. Consider, say, Darwin's theory of the origin of the species, Marx's theory of the class struggle, Freud's psychoanalysis and Saussure's theory of language. Not even all physical theories have proved axiomatizable – and certainly not on the positivists' conception. Nor do models have a role in such an axiomatization. And lastly, the proposal is unable to explain the evolution of theory over long stretches of time, or the particular fault lines that appear when one theory is replaced by another.

The syntactic position sees linguistic meaning as generated from the bottom up. By contrast, the *semantic* position sees it as a top-down process. The first formulation of the semantic position was offered by Patrick Suppes, around 1960. Since then it has gained many adherents, among them Bas van Fraassen, Frederick Suppe and Ronald Giere. The central idea was taken from formal model theory. Its core claim is that a model is an interpretation of a theory that makes the theory true, and that a scientific theory consists of a set of models.

A scientific theory is a formal, axiomatized structure that can be given a semantic interpretation provided that it satisfies the following series of abstract and formal conditions:

(i) The theory T must be formulated in a first- or second-order language L.
(ii) The language L can be given an interpretation I by postulating that the individual variables run over a domain of objects and that predicate variables run over a series of values and properties that satisfy the

predicates. An interpretation is the assignment of a class of objects and properties that will satisfy the axioms and theorems in L.
(iii) Every sound interpretation *I* that makes *L*'s sentences true with respect to the postulated entities is a model of *L*.
(iv) A theory is identical with an entire class of models that are each individually a consistent interpretation of *L*.
(v) The set of models, a family of models, constitutes the truth conditions for the theory, *i.e.* all the possible interpretations that invest the theory with truth and falsity.
(vi) If one of these models represents the real world the theory is true with respect to the actual facts.

We can summarize this semantic interpretation of scientific theories under four points: 1) a theory consists of a set of interpreted sentences; 2) a theory is true or false; 3) a theory does not refer directly to physical objects and their properties, the connection between the set of interpreted sentences and the world is mediated by a model; and 4) a model consists of non-linguistic, abstract entities.

The semantic position has it, then, that an axiomatic system acquires semantic meaning through a formal interpretation on which to each logical symbol is assigned an abstract object or property in the model. But here problems begin to emerge, for every interpretation is formed on the basis of a prior understanding, and every interpretation must necessarily build on the linguistic meaning we already grasp. According to the semantic position, it is not the relevant scientific practice that contributes to the meaning of axioms and theorems. The abstract objects and properties do that. So it would appear that we learn the meaning of a theory by being 'acquainted' with these abstract entities and by knowing when they make the theory true and when they make it false. But the claim that we understand a theory's meaning because we are immediately 'acquainted' with such independent abstract objects and properties is far from compelling. For what affords us access to them? It would seem highly improbable that we should be better equipped to apprehend them than we are the physical world, which we can observe with or without the aid of instruments.

The semantic position has other defects too, matching those we identified in the syntactic position. We can scarcely expect to be able to formalize many scientific theories other than those in physics. Once we turn to cosmology and chemistry, it becomes difficult to elaborate a semantic interpretation.

Both the syntactic and semantic interpretations regard theories as true or false, or as approximations to truth or falsity. Notably Bas van Fraassen has argued that the epistemic goal of science should not be truth but empirical adequacy, with empirical adequacy construed as the isomorphism obtaining between a semantic model's substructure and observational sentences. While subscribing to the claim that scientific theories are true or false, he denies that we do or ever could have methods to establish their truth-value. He contends that even the most reliable methods are inadequate to the task of fulfilling an epistemic goal such as truth.

The semantic interpretation is unsatisfactory because it assigns to abstract entities the role of making a theory true or false; it is these that we need to be acquainted with if we are to understand a theory's import. The position has the semantic contents of theories percolating down to scientific practice without this same practice having any influence over how the meaning of the scientific theories is to be understood. Only the *linguistic* position allows for that.

The linguistic interpretation marks itself off from the two others by depriving theories of any claim to truth content – theories are not susceptible of truth or falsity. Theories provide us with a vocabulary and a set of linguistic rules that govern its use. What *does* qualify as true or false are *scientific explanations* in the shape of concrete descriptions of reality. We do not apply our theories directly to reality, but to those abstract entities and relations the models represent in order to describe a particular process and, by so doing, to explain the phenomena represented by the model. The theories dictate, as it were, the framework for the articulation of models that are idealized representations of reality. By reference to models we produce explanations of concrete problems, and it is only such individual explanations that are true or false.

But what does it mean to say that theories consist of a set of linguistic rules and a vocabulary? Explicit linguistic rules normally lay down the use of the vocabulary by defining the words and definitions as analytical determinations of how particular expressions are to be understood. But are definitions not the result of some stipulation, as when we decide that 'bachelor' and 'unmarried man' mean the same thing? Is an expression like 'Force is the same as mass times acceleration' not also the result of stipulation? Nothing could be further from the truth. Meaning does not come from above, nor merely from below. We should regard definitions as a kind of analytic *a posteriori* statements. Experience has had no input into our definition of the word 'bachelor'. By introducing this word, we have, in the interest of linguistic simplification, merely made it possible for one to do the work of a complex of words. Naturally, such words can occur in a theory too. But Newton's first law gives us a new definition of *natural motion*, *i.e.* motion that needs no explanation – a definition at odds with Aristotle's. The law of inertia is a rule-conferring definition that lays down the stipulation that henceforth natural motion is rectilinear. Newton's second law similarly yields a new definition of the concept of force – a concept we already possessed – and expresses it in terms of concepts that were also already familiar. Aristotle's own concept of force had its roots in the power familiar to human *experience* – from the need to raise a boulder, bend a bow and draw a cart. Such experience delivers, if you will, the word's primary *referential meaning*. This meaning is then gradually extended through the introduction of new criteria that specify the standard operations for measuring force. Via definitional linkages with the concepts 'mass' and 'acceleration' the concept of force receives its *theoretical meaning*. Starting with its primordial referential meaning there accrues to the concept, first with Aristotle and later with Newton, an increasingly theoretical content that contributes to the word's ideal extension. The theoretical meaning, however, is susceptible of modification if experience shows that allegiance to the model shifts, thereby causing the need for a revision of theoretical meaning. So what happened in the move from Aristotle's theory to Newton's was that science's commitment to a model (or a family of models) changed.

The difference between theory, model and explanation is, ultimately, the same as that between natural language and actual descriptions formulated in the language. Theories are not concrete descriptions of concrete individuals in the world but models open up for such descriptions. Every natural language functions in virtue of syntactic and semantic rules for the language in question. These linguistic rules are naturally neither true nor false. They determine what is normatively correct if a description of the world is to be consistent and coherent. Natural language only begins to be something that can be true or false when it serves as the vehicle for individual statements about *concrete* facts: things, properties, relations, situations, processes and circumstances. The only difference is that scientific theories constitute a considerably more nearly defined set of linguistic rules than are normally met with in natural language because they do not treat actual, but idealized things. They also specify norms for what is correct to say, and what not, if there is to be coherence and consistency. The function of scientific models is to yield an abstract and idealized representation of particular events and processes which – given a particular theory – enable us to frame particular, individual statements about such concrete phenomena, and these statements will be true or false.

An example will elucidate this. James Clark Maxwell's theory of electromagnetism furnishes scientists with a language in which to talk about the electric and magnetic properties of all kinds of phenomena. On the surface of the sun sunspots form which reflect the sun's magnetic processes. These sunspots are most active in 11-year periods, but the length of the periods varies from just under 10 years to 11.5 years. The more intense the solar activity, the shorter is this interval. Around 1990 two Danish scientists, E. Christensen and K. Lassen, discovered that there was a clear correlation between this variation in the frequency of sunspots and the average temperature of the earth. But such a correlation does not necessarily imply a connection. In the absence of a causal story it might equally well be a coincidence. Six years later, however, another Danish scientist produced just such a story. H. Svensmark put forward the idea that cloud formation might be influenced by cosmic radiation that was in turn under the influence of the magnetic field of the sun. He framed a theoretical model of the relations between sunspots, the magnetic activity of the sun, cosmic radiation, cloud formation and the warming of the atmosphere near the surface of the earth. His model was based on already familiar explanations in physics, each of which rested on other tried-and-tested models and principles in physics, thereby enabling him to formulate his causal story. It runs as follows: The magnetic field of the sun shields our solar system, including the earth, from cosmic radiation. Intense magnetic activity reduces radiation; cloud formation is correspondingly reduced and leads to an increase in the warming of the earth's surface. Conversely, weak magnetic activity produces increased radiation that results in greater cloud formation and the warming is reduced. It is this explanation of the correlation between sunspots and average temperature that is either true or false.

Now it might seem reasonable to ask how it could be possible for theories to be used to predict new phenomena if they are neither true nor false. Underlying such a question is the conviction that theories can be used to predict wholly new phenomena only if they are true – otherwise not. In my view this reasoning rests on a misunderstanding.

There is of course a difference between just talking about novelties and actually predicting them, but we are in fact able to make predictions in everyday life. I can predict that I will fall off my bike shortly after ceasing to pedal and that a flowerpot, falling from the fifth floor, will hit the pavement and break on impact, and that the next time I spot a dog in the street it will have something in the order of four legs, a head, body and tail. There is no difference between predictions of this kind and many scientific ones. The discoveries of Uranus and Neptune, in 1781 and 1846 respectively, were of course based on the observation that Jupiter's orbit of the sun is influenced by that of Saturn and that, first, Saturn's trajectory – and then Uranus's – did not correspond to that described by Newtonian mechanics as applied to a two-particle model. It is experience that enables us to make new discoveries, not theories. If we believe it to be the other way around it is because experience is always already conceptualized or expressed in terms of a model when it leads us to new discoveries.

But there is another side to this that needs to be considered. What might well look like predictions are in fact possibilities inherent in language. A language has properties and structures quite independent of whether the individual language user is aware of them. It has often been shown that theories have implications that their proponents only gradually came to recognize, but which must have existed already at the theory's inception in virtue of being intrinsic to the language in which it was framed. When such possibilities are developed they take on the character of fresh predictions. For example, Poisson was able to show that Fresnel's wave theory of light implied the existence of a bright spot at the centre of the obverse of an illuminated disc. What happened was that the theory was applied to a concrete model representing the movement of light around an illuminated disc, and it was the model and not the language that proved capable of explaining the spot's emergence.

Natural language can also be used to talk about novel phenomena: it affords us a wealth of ways in which to describe entities and their properties. Natural language is so rich in resources that we can often talk about new things without needing to introduce new terms to designate them. But if we need to, we do so, and are thus continually expanding our stock of terms. We can describe a unicorn without its needing to exist, and should exemplars be found, this would not have been predicted in virtue of the mere existence of the word: language, as such, is not 'true'. By the same token, scientific theories are full of semantic structures that do not fit any concrete model and thus cannot be used to explain aspects of reality. Perhaps there are no such things as magnetic monopoles, Higgs particles, negative energies or advanced potentials, even though physical theories make it possible for us to talk about such things.

We have yet to give an account of how the linguistic position views the semantic meaning of theories, but before turning to that issue we need to address the function of models in science.

Models

Models are pivotal to science. They mediate the connection between theory and reality and enable the scientist to devise new theories on the basis of what has been gleaned from experience. There are different kinds of scientific models – both physical and theoretical. One of the differences between the semantic and the linguistic interpretations comes out in their respective understanding of models. A theoretical model should not be construed in model-conceptual terms merely as an interpretation of the theory, but as a bridge between both theory and reality – an interpretation of theory and an idealized representation of reality.

The difference between theories and models may be characterized as follows. A theory does not define things in terms of their properties but merely introduces properties and defines their interconnections; a model uses these concepts to represent things and the causal connections between them. Recall Newton's three laws. They have to do with the properties of mass, velocity, acceleration and force, and have nothing at all to say about planets, pendulums or tides. The role of theories is thus to introduce and fix properties while the role of models is to represent concrete entities with properties which are such that they satisfy the laws of the theory, and in a manner that takes account of the *ceteris paribus* clauses which are imposed on laws when we idealize concrete things and prescind from the actual situation.

A theoretical model is always a representation of a concrete reality on which the selected items are represented by idealized and abstract objects and *ceteris paribus* laws. We can, for instance, represent the earth and the sun in a theoretical-particle model in which they are mapped as point masses with a central force operative between them. The model can naturally be made more or less exact depending on the extent to which we take into account the system, the physical circumstances applying in the particular case, and the specific epistemic aspirations and interests we might have in explaining some aspect of the behaviour of the system in a particular way. When the theory is applied to the model, the scientist is able to offer an explanation of a concrete phenomenon.

Take, for instance, a container with gas in it that we want to heat to a certain temperature. Prior to so doing we want to know the pressure the sides of the container will be subjected to in the process. We are not interested in all the particular gaseous properties. Those on which we focus will have been selected relative to our epistemic interests. We know that a gas has a temperature T, pressure P and volume V and that the relation between these properties is expressed in the general equation of state also known as Boyle-Mariotte's law: $PV = nRT$. These properties are ones we can measure. And even if Boyle-Mariotte's law, in line with all phenomenological laws, is limited by *ceteris paribus* clauses, within the range of moderate temperatures, it is capable of telling us how great the critical pressure must be.

If we are interested in understanding what causes the properties to impact on each other, and therefore in the temperature range outside of which Boyle-Mariotte's law does not apply, we are not interested in what other physical and chemical properties the gas might have and not at all in its taste, odour or colour. We conceive of the gas as composed of a mass of invisible particles, and so we need to have the temperature, pressure and volume represented by the mechanical properties of these particles. Gas molecules are assigned to properties such as mass, velocity, elasticity and the ability to collide with each other and with the walls of the container, and the temperature is considered to be a function of the average velocity of the many particles. Again, we are not interested in all the properties of particles. We select individual ones and idealize them. This results in two assumptions about the molecules: 1) no force is operative among them; 2) they have no extension. We have thus devised a model of an *ideal* gas in the container. We are now able, on the basis of linguistic rules fixed by the kinetic theory of gas, to derive Avogadro's law, the equation of state of an ideal gas, which specifies how such an ideal gas will behave, and we shall thus have provided a causal explanation of the actual pressure at a given temperature. Avogadro's law is a *ceteris paribus* law as well as a fundamental law.

A real gas is in no way an ideal gas. We may discover that the model is insufficiently exact for our purposes when we map particles without structure and extension. For it to be able to yield a fair reproduction, the molecules have to be without extension and the distance separating them from each other has to be relatively great. If this is not the case, it will be necessary to investigate whether the gas in the container is monoatomic or multiatomic. Even this distinction may not be adequate to the task, and we will need to add more properties and structure to the model. We must take account of what kind of gas is in the container, the mutual attraction exerted by the molecules and each individual molecule's volume. The kinetic theory of gas now gives us van der Waal's equation of state.

All three equations of state express *ceteris paribus* laws which do not apply to any actual gas but to ideal gases in a model. No single mathematical expression, no individual law, would be capable of describing actual gas. The model affords the scientist a means of interpreting and representing reality and it is to this abstract structure that she applies her mathematical theories and develops new ones.

We construct models also within the social and human sciences either from the ground up or on the basis of those theories we have already developed within all these areas of knowledge. Such models reproduce persons, groups, organizations, nations, languages, artworks, religions, etc. as abstract entities with ideal properties. Models make it possible, for instance, to explain an action on the basis of its intended purpose or a work's meaning and structure on the basis of the narrative intentions embedded in it. We create a model of a novel when we assume that the characters in it are, as are real people, the bearers of particular characteristics. Some of these characteristics consist in unconscious desires and ideas, drives and conflicts that to a greater or lesser extent may be repressed. Such complexes come to expression in the individual's acts and utterances as well as in those physical objects in their environment which might symbolize these unconscious desires and dreams.

Let us take a closer look at the basis for a psychoanalytic interpretation of people's unreflective actions. The standard model for psychoanalysis partitions the individual's mental life into consciousness and the subconscious and represents all the psychic states that surface in the conscious mind as accompanied by pleasure or aversion. States marked by aversion are subjected to *repression* (defence mechanisms) to the subconscious only to reassert themselves in the conscious mind as *symbolic distortions* in the form of neuroses, dreams, slips of the tongue and other actions. To describe these elements and their interactions in this model the researcher uses psychoanalytic theory. Like any other, the theory contains a vocabulary and certain rules for its use:

1. *Topical vocabulary* The psyche consists of a number of elements: Superego, ego and id. The superego = the censor, conscience; ego = consciousness; the id = drives and unconscious desires.

2. *The economic linguistic rule and vocabulary* The stimulation which the psyche receives from its drives: a) the instinct of self-preservation (life drive), b) the sexual drive (*libido*) and c) the death drive (*thanatos*).
3. *Dynamic linguistic rule* Psychic conflicts are the result of conflicts that derive from 1) and 2).

The model may be further expanded through the introduction of other theoretical components:

4. Infantile sexuality.
5. The topical unconscious (the id.).
6. The Oedipus complex constitutes the core of the unconscious.

One method of uncovering unconscious structures in the mind is through an analysis of the individual's actions. The method is commonly applied to dreams, which are regarded as symbols, and whose occurrence is produced by displacement (metaphor) and compression (metonymy).

This kind of model makes it possible for us to explain a person's dreams which would otherwise defy our attempts at understanding. As already noted, psychoanalysis is, therefore, not a true or false theory but one that provides us with certain rubrics, the use of which enables us to describe the actions of the characters in a novel so as to render them intelligible.

We might also have chosen a different model of the novel according to which its narrative sequence is seen as a structure with a particular content. On this alternative approach we assume that the individual's character traits and actions are the expression of a sequence and a structure which are combined in a project – Greimas' so-called 'actant'-model according to which the individual's characteristics and actions evolve via serial phases until the project is achieved. This model makes it possible for us to use structuralism's theory of the textual analysis of patterns and semantic oppositions, positive and negative values, etc. Again, we have a theory which is not in itself true or false but which opens up avenues leading to an understanding the text, once a model for it has been set up.

These examples of the relationship between scientific theories and models give rise to several questions: 1) how do theory, model, explanation and reality hang together? And 2) how does the answer to 1) relate to the philosophical discussion of realism and antirealism? We shall address the latter in the next chapter. Here we shall concern ourselves with the former.

Models are no more true or false than are theories – even though for other reasons. Models represent real things: they are not linguistic objects but abstract or physical structures that are assumed to bear certain resemblances to real phenomena. Explanations, by contrast, may be true or false. Take, for example, a map of Rome: it is a physical representation of the city but not a representation that is true or false. It shows certain features of Rome while omitting many others. It is a partial representation showing us the interrelationships of the streets but not the distances between buildings and squares. Famously, Rome is built on seven hills and the map is a two-dimensional representation. So it depicts Rome by virtue of reflecting the spatial configuration of its parts. Everything east of the Tiber is on the right of the river as depicted on the map, and what is west of it is on the left. The depiction can be good or bad. But it is what we say about Rome on studying the map that is true or false.

Models can be used to articulate concepts and construct new theories and they may be used to produce fresh hypotheses and explanations. Look at Figure 1. It shows us a model of the function of theoretical models in science. A series of arrows connects the various components. Each arrow stands for an entry or an exit individually representive of a particular practice that connects one component with another. The connections go in both directions: from reality to theory and from theory to reality. It is, in the first instance, the particular scientific practice as underpinned by observations, experiments or readings and governed by methodological rules that help us to create a model of reality. The model makes it possible to develop new theories and frame explanations. But theories may also be applied when models are described, and these descriptions can be used to tell a story which may be either tested or in some other way confronted with reality. Is there a fit, or not? The fitting of model to reality is a dynamic two-way process.

Let us illustrate this by means of a concrete example, the model of the atom. In 1907 Ernest Rutherford discovered that he could shoot alpha particles through thin gold foil without most of them rebounding back at him. A few did; others were deflected in various directions, but most of them went straight through. This led him to construct a model. He pictured the atom as a small solar system. Given that the alpha particles were positive and were repelled by positive particles there had to be a massive positive nucleus at the centre of the atom which, in order for the whole atomic system to appear electrically neutral, needed to be surrounded by negative electrons that orbited the nucleus at some distance from it (relative

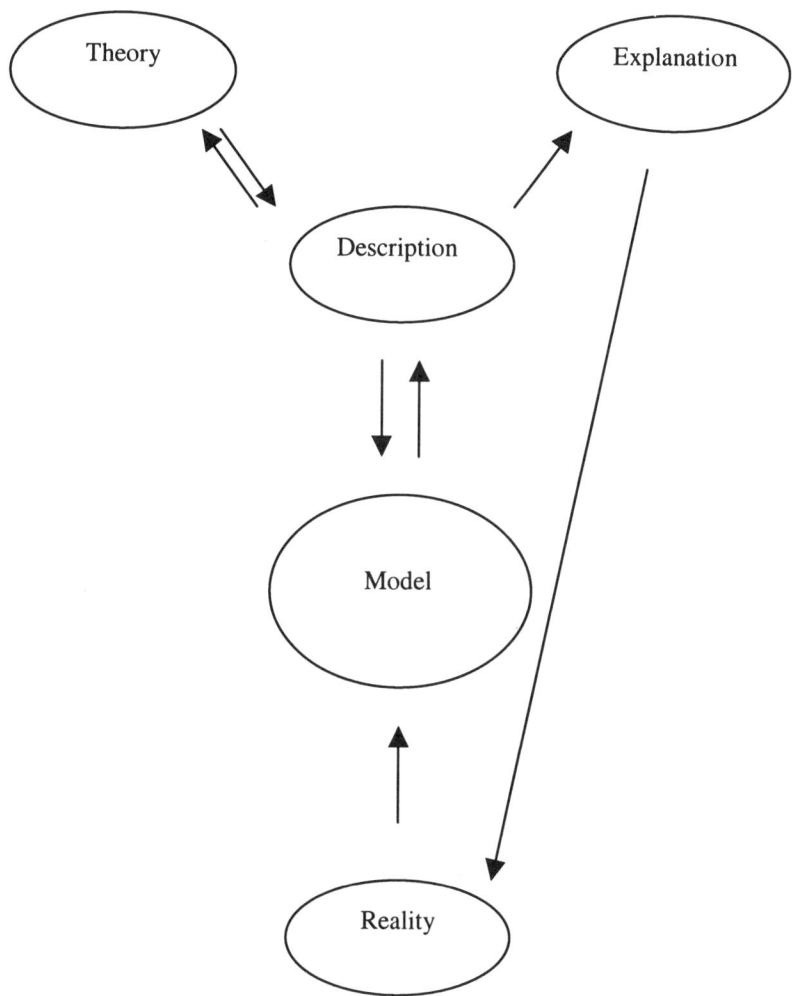

Figure 1 Flow diagram of identifications and interpretations between different elements of the scientific practice

to the size of the atom that is). But the model was not without its problems. When classical electrodynamics was applied to it, it soon became clear that it delivered inconsistent answers. The electrons would be attracted by the nucleus, and when that happened the atom would collapse. Niels Bohr retained the model but modified it; thereby paving the way for the development of a new theory that would come to be called quantum mechanics. On Bohr's model, as long as they were bound to the atom, the electrons could move only in particular orbits around the nucleus in virtue of having quantized energy, and they are not located at all at points in between. They could, however, jump from one orbit to another – either as the result of being jostled by other particles or spontaneously emitting energy with a certain wavelength. For a long time Bohr's model functioned in the absence of a theory. It was used to explain experimental results and these would deliver feedback when the explanation was insufficiently precise. In this way Bohr, and others, produced serially improved versions of the model. Understanding of the atom progressed apace, culminating in 1925 with Heisenberg's, and in 1926 with Schrödinger's, formulation of quantum mechanics. A crucial guiding principle in the progression from model to theory was Bohr's methodological and semantic constraint on any putative theory to the effect that its description of the transition from the bound to the free electron should correspond to that of a free electron in classical physics.

But where does semantic meaning originate? In theory or in practice? In both. According to the linguistic position it is not a case of 'either/or' as it is on the syntactic or the semantic positions. Scientific practice also plays a part in the construction of scientific concepts. Reality comprises natural or nominal objects which appear or behave in a specific way, and which we seek to explain. Their sortal properties are ones we can identify in our practical dealings with real things. Recall the criterial theory of meaning. The practical aspects of our identification of things include assigning certain criteria to them, and such criteria enter into the meanings of the words we use to designate them. This applies to the case of properties as well. Take mass, for instance. We have different practical means of establishing the mass of particular items, and these determine the reference of the word and thus become part of the word's referential meaning. In a model the idealized properties of the abstract objects are supposed to represent factual properties of real-world objects, and the model's properties partially determine the semantic meaning in the language by which we describe the model.

The model is not itself a linguistic system but a system of abstract objects describable in a language defined by the theory. The physical notion of mass – resulting from empirical and experimental practice – is represented in the model as an idealized property of these abstract objects; the same notion is then connected up with the ideal expressions of the theory through the stipulative identification of its reference in the concrete description with the idealized property of the model. One way of delivering a more detailed account of the way in which we use the theory to describe a specific model would be to view the relevant stipulation as a semantic model of the theory, which we then identify with the models of the concrete system which we have abstracted from experience. The meaning of the word is determined by the rules that the theory specifies for its use in describing the model. First the physical concept is linked to the mathematical symbol 'm'; then, the theories determine the rules governing the symbol's inclusion in a dynamic description of the model whose objects have the properties that are fused in the theory. So part of the word's meaning comes from practice and part from theory.

The specific interconnections between theories, *ceteris paribus* laws, models, and reality are made up of many complex and heterogeneous practices. We can establish that scientists use methods of discovery to frame models, some of which, over time, attain the status of standard models. For a standard model the *ceteris paribus* laws involved are well known (be they the causal laws of natural science or the functional, intentional or interpretive laws of the humanities); when the explanation of a particular phenomenon is attempted, the phenomenon in question will figure as a member of a kind, as an abstract object, thus becoming describable through a particular theory or our general linguistic practice. Furthermore, the model is adjusted and fine-tuned to fit the relevant phenomenon, so that so many as possible of the specific circumstances that are deemed relevant are expressed in the model.

Paradigms

But science is more than just theories and models. It includes, moreover, metaphysical conceptions and ramified practices, which enter into the development of new theories and models. For many years philosophers of science were interested in understanding the nature of scientific theories but gave scant attention to the practical aspects of experiments, observations, explanations and interpretations that contribute to the characterization of science as a particular epistemic sphere that is an extension of our lifeworld. Scientific understanding does not rest solely upon a theoretical analysis of phenomena, but also to an equal extent it replies upon the practical interaction of the scientist with phenomena. A theory cannot attach itself to reality; that can occur only through scientific practice fixing the meaning of the theories in virtue how we use them. The connection between thought and world is mediated by agency – it is via agency that language gets to represent reality.

The year 1962 saw the publication of Thomas S. Kuhn's book, *The Structure of Scientific Revolutions* in which he describes the historical evolution of science. What was new about this description was, among other things, Kuhn's treatment of metaphysical conceptions as an important key to the understanding of the scientific process. Science, says Kuhn, is made up of paradigms, and usually just one paradigm governs each scientific discipline. A paradigm is a suite of positions shared by a group of scientists which establishes consensus among them, directs their research towards the resolution of new problems, and determines their conception of reality. The progression of science is characterized by successive revolutions with the prevailing paradigm, once it has proved itself inadequate to a range of intractable problems, being replaced by another.

Kuhn sees each natural science evolving through four phases: First there is the *pre-scientific* epoch with no unifying paradigm. This period is marked by the absence of any consensus among scientists. Each researcher has his peculiar theory. No explanation may be identified as superior to another. So long as the scientists' conceptions are not subsumable under one paradigm, there is no activity being pursued that we can call mature science. Its emergence occurs at the point at which a group reaches agreement on a shared paradigm – the condition for the development of mature science as such. Research, then, enters the phase of *normal science* in which scientific activity is directed by the application of the paradigm to concrete puzzles posed by various phenomena falling under that paradigm's scope. It must be developed and adjusted, articulated and adapted, so that over time it becomes applicable to more and more phenomena. Kuhn calls this activity puzzle-solving and he compares it to the solution of a crossword or a jigsaw puzzle. The paradigm delimits the range of admissible solutions and nature returns an answer in each individual case. Normally, the paradigm admits only one answer, just as the crossword admits only one word and the jigsaw one piece. Should the scientist meet an intractable phenomenon which does not immediately harmonize with the solution offered by the paradigm, she does not regard the result as a falsification of the paradigm, but rather as a test of her own ability to grasp the phenomenon in light of the theory. If the phenomenon continues to be recalcitrant we have an *anomaly*, and if such anomalies multiply the result will be a *crisis*. The time is now ripe for alternative proposals to gain a hearing among scientists. It may be that the prevailing paradigm will retain the ascendancy, or perhaps the new paradigm will win support. In the latter case science will have undergone a revolution.

Initially, Kuhn's characterization of a paradigm lacked precision. The concept was blanketly applied to highly specific theories, general theories and metaphysical assumptions. He was later to give it a more precise cast. Paradigms comprise four elements:

(a) metaphysical presuppositions,
(b) symbolic generalizations,
(c) exemplars, and
(d) values and standards.

Second, metaphysical assumptions are a medley. We have already encountered them in the form of ontological principles. Claims may be advanced to the effect that nature is or is not deterministic, that nature abhors or does not abhor a vacuum, or that nothing can come from nothing or the negation of that contention. In point of fact, we also find ontological principles in the human sciences: that human subjects do or do not have a free will, that consciousness can or cannot be reduced to neuro-physiological processes in the brain, that linguistic meaning is more than or, alternatively, just *is* behaviour. What characterizes such assumptions is the higher-order role they play for thought. They often enter in as tacit presuppositions underlying every scientific explanation.

Another component is constituted by symbolic generalizations. They commonly constitute the core of a scientific theory. In classical mechanics, for example, it is Newton's second law, $F = ma$ which says that force is equal to mass times acceleration; in electricity theory it might be Ohm's law, $I = V/R$ which says that electric current is equal to electric tension divided by resistance. According to Kuhn, such regularities viewed in isolation are of limited application. It is only in the concrete context that the scientist can use them and give a precise description of mechanical or electric systems.

Here exemplars enter in as a third element. They represent a kind of ideal model for concrete answers that students meet in the course of their training, in laboratories, in exam colloquia and in the exercises at the end of scientific textbooks. More technical solutions also encountered in the scientific literature may serve as a template for the researcher's solving of concrete problems. Kuhn regards such exemplars as forming the principal plank of the paradigm. In the social sciences, and the humanities as well, we find exemplars that enable the student to apply specific conceptual apparatus. It might be the analysis of a particular economic situation, of a political or social configuration of factors, or the interpretation of particular works. Freud's psychoanalytic interpretation of Irma's dream, and Marie Hochmuth Nichols's neo-Aristotelian analysis of Lincoln's first inaugural speech, are instances of such exemplars in the human sciences.

Finally, the fourth element to be encountered is the relevant values. They include methodological and epistemic values that do not necessarily belong to a single paradigm but are prescriptions one may have in common with others. The most preponderant values are comprised by the constraint that the paradigm be capable of explaining observed phenomena, and of leading to empirically ascertainable predictions. But, equally, the values might be more specific and include, for example, precision, simplicity and consistency. Many of these values do not really belong to one particular paradigm; it is more accurate to say that different paradigms will exemplify them in varying degrees and ways. They have less to do with constraints on an individual paradigm than with constraints applicable to all sciences.

A basic feature of the paradigm is its *incommensurability* with other paradigms. Each paradigm determines its own semantics and in consequence there will be semantic differences between them. Put otherwise: two paradigms are incommensurable if they cannot be translated one into the other without significant semantic loss. The claim of significant semantic incommensurability is grounded in a holistic theory of meaning according to which a word's meaning is determined by its interrelations in the entire web of linguistic meaning determined by the language as a whole. Identical expressions occurring in various linguistic systems may accordingly bear different meanings. Among other things, paradigms are discrete linguistic systems and the same expressions will thus not have the same meaning in all. For instance, 'mass' in Newton's mechanics and Einstein's relativity theory respectively will mean two different things. This creates problems for paradigm choice, for the scientist is barred from appeal to experience since the very description of scientific observations is paradigm-dependent. Consequently, the choice between rival paradigms cannot be made on the basis of a rational method because each theory carries different semantic content. Kuhn compares the choice to a Gestalt shift in which the description of an experience in its entirety changes as we move from one paradigm to another. What the scientist sees varies depending on whether he describes the world on the basis of Newton's physics or on that of Einstein.

What is attractive about Kuhn's paradigm is the comprehensive manner in which natural science is identified. Natural science does not consist simply of symbolic generalizations but in equal measure of metaphysical assumptions, ideals, values and prescriptions. And the concept of paradigm is enough broad to be expanded to embrace all science. Symbolic generalizations that are otherwise often identified with a formal scientific theory are not simply to be regarded as mathematical equations. Rather, they constitute general conceptual structures that are expressed in explicit rules of language, irrespective of whether they are formulated in mathematical notation or ordinary language. Metaphysical assumptions and models, as well as values and standards, render this concept equally applicable in the characterization of the human sciences. The idea of paradigm shift and competition between alternative theories has thoroughly pervaded the social and human sciences and taken root there. The problem with Kuhn's analysis of science, however, is the notion of paradigmatic incommensurability. He has emphasized too radically the semantic variance of paradigms. In Kuhn's presentation of the paradigm there is, oddly enough, very little reference to scientific practice. This feature influences his conception of a scientific paradigm and his view of the choice between different paradigms.

The progress of science emerges in Kuhn as a non-cumulative process. Scientific knowledge is not a case of building with building blocks by adding fact to already accumulated fact. What science deems knowledge is a function of the prevailing paradigm. Once it was a presupposition that the sun orbited the earth but no longer. By the same token, it was once a fact that leeching was a cure for diseases but not any more. And the time was when it was believed that Homer was the unique author of both *The Iliad* and *The Odyssey*, but today we know better. Many will reply that these implications are plainly unacceptable. For starters, they will point to the difference between a fact and the *belief* that something is a fact. We cannot have knowledge of a fact if it is not really a fact. The belief that the earth goes round the sun does not entail that it does. Indeed, we can say that physics abandoned the geocentric paradigm precisely because what was believed proved not to be a fact. Kuhn is only right if paradigms are incommensurable and if paradigms determine what is and what is not a fact. On that scenario, the choice between them cannot be made by reference to actual circumstances.

Kuhn is ambiguous on this point. On the one hand, he says that standards and criteria of evaluation are only in part paradigm-dependent and that that, to a certain extent, makes a rational choice between alternative paradigms possible. On the other, he stresses that there are no paradigm-independent facts because all facts get described within a paradigm, and we cannot translate from one to the other without residue. It is not difficult when reading Kuhn to get the impression that the world changes whenever there is a paradigm shift. All the same, close attention to the text reveals that he never claims that the world has changed but that our conception of it has. Scientists see different things but the world remains the same. For that reason it is infelicitous to say that facts are paradigm-dependent – it is at most the scientist's belief that a particular statement denotes a fact, that is contingent upon a choice of paradigm. Facts are facts regardless of whether they are believed to be such.

Since different paradigms give rise to differing descriptions there is no need to follow Kuhn all the way. The choice between alternative paradigms does not rest exclusively on possible common methodological and epistemic values. Kuhn fails to take account of the fact that there is a common scientific practice that cuts across paradigms and makes it possible for that scientists to communicate with each other. A shared basis of experience arises through the scientist's engagement with the world and his practical work of identifying natural kinds and nominal artefacts, and such a practice can be, to a considerable extent, freestanding in relation to the prevailing paradigm. In what is a combined observational, experimental and linguistic practice the scientist is inducted into the 'correct' use of names for natural and nominal objects. Language acquisition in this case occurs not via the theory, but by the observational and operational criteria which fix the use of names entering in as an analytical *a posteriori* determination of the reference of expressions. This means that a scientist can go on talking about the atom and use the name to refer to one and the same natural kind irrespective of whether the properties figuring in the model change and of whether the model in question is Thomsen's, Rutherford's, Bohr's, Heisenberg's or that of another.

There is an alternative to Kuhn's approach to scientific evolution. The Hungarian-born philosopher Imre Lakatos developed what he called the methodology of research programmes. He was dissatisfied with Kuhn's paradigms because that line of thought has scientists incapable of making a rational choice between different conceptual schemes, thus ruling out scientific progress. However, like Kuhn, he sees scientific evolution as a struggle between competing theories rather than a confrontation between individual theories and naturally given facts. As with Kuhn, science is more than a set of regularities: it is a system of metaphysical assumptions, theories and methodological rules, which Lakatos calls *research programmes*. And like Kuhn, he does not think that experience can falsify a theory in the absence of a replacement for it.

A research programme consists of a *hard core* surrounded by a *protective* belt. The hard core defines the research programme that constitutes the foundation of any theoretical system. It might be Newton's three laws of motion, Schrödinger's wave equation, Darwin's law of natural selection, Marx's economic laws or Freud's doctrine of Oedipal repression. Lakatos held that this kernel identifies a field of research, and that no scientist is interested in abandoning the core since such abandonment would close off the possibility of understanding the phenomena. Scientists would normally regard the kernel as immune to direct viral attacks from anomalies. The immune system consists of a protective belt of assumptions and auxiliary hypotheses which, in conjunction with the hard core, are needed to explain the observed phenomena; but it is continually open to the scientist to question and alter it. It will include such things as assumptions about the concrete conditions surrounding the item under investigation, external influences, or auxiliary hypotheses about the function and reliability of the apparatus in question, methods for the collection of data, credibility of sources, etc.

The research programme contains further methodological rules in the form of a *negative heuristic* and a *positive heuristic*. The negative heuristic briefly states that the hard core must not be abandoned in the face of anomalies: it should be protected from falsification. The positive heuristic drives the research programme forward; it offers a strategy for action in the face of anomalies that the research programme cannot explain, and for the further development of the programme.

It is in what concerns the transition from one paradigm or one research programme to another that Kuhn and Lakatos really part company. For Lakatos sees this transition unfolding in two ways: The research programme may be successful, develop *progressively*, when it is better than other programmes at accounting for new as well as old observations. As long as the protective belt can neutralize possible anomalies, because it can be modified or replaced, there will be no rational ground for giving up the hard core. But the programme may also end in failure. If the programme increasingly draws on *ad hoc* hypotheses to protect itself, this spells the beginning of the end. It stagnates and, no longer able to explain fresh phenomena, it degenerates. In such cases it is rational to abandon it in favour of another programme.

A progressive research programme exhibits the following features:

(a) the possession of content not belonging to earlier programmes;
(b) the capacity to explain why earlier programmes were successful; and
(c) its confirmations are better than those of earlier programmes.

On the face of it, the constraints on a progressive research programme would seem to dovetail with the constraints on an inference to the best explanation. The research programme that best explains most phenomena is the most successful. But we have also noted that hypotheses may be empirically underdetermined and it is therefore not possible to decide between competing hypotheses on the basis of experience. Paradigms and research programmes are no exception. We may find ourselves in a situation in which we decide between different paradigms, but where we cannot make a rational choice. It all depends on the scientist's particular taste.

Lakatos is constrained to give the same answer to the question of when the scientist should abandon the sinking ship. For even if the incumbent programme is leaky, it will often still be superior to and more developed than a new and as yet unelaborated programme. And to the question of how we can explain that an alternative programme will be able to fulfil conditions (a)–(c) without appeal to the predilections of the individual scientist, Lakatos replies that it rests on a methodological convention. The choice, then, builds on a subjective judgement and it is only retrospectively – when the programme proves capable of explaining new subjects – that the choice is seen as progressive. There is nothing exceptionable about that. All rationality builds on conventions. But such an answer brings Lakatos disturbingly close to Kuhn who also holds that the choice of paradigm is grounded in the scientist's social predilections. Not, admittedly, merely personal and untutored predilections, but a well-considered proposal that can both be discussed and grounded by reference to methodological values. The rationality of choice is simply an expression of how many are prepared to subscribe to the values in which the choice is grounded.

The disagreement between Kuhn and Lakatos turns on the question of whether it is possible to make a rational choice from among different conceptual schemes. What Kuhn sees as the choice between incommensurable paradigms, Lakatos sees as one between commensurable research programmes. But behind the verbal formulations we find that for both the choice is made with reference to particular values and conventions, which can of course be appraised and ranked variously. But what is interesting is the fact that neither holds that the choice is grounded in an appeal to the truth of a theory. The move is not from a less true to a more true theory – a revolution or progressive transmutation of a problem does not give us more of the truth. For both theorists, the constraint on a scientific theory is that it is empirically adequate – not that it be true (or closer to it.) The constraint is, accordingly, not far removed from the position which maintains that Newton's theory is, as little as is Einstein's, the bearer of truth. For if scientific theories are neither true nor false, successive theories cannot be held to represent increasingly closer approximations to truth. A theory consists of a lexicon and a set of linguistic rules evolved to enable discourse about a particular domain of objects; but over time its successful use may prove to be more restricted than originally thought.

References

Giere, R. (1988), *Explaining Science*, University of Chicago Press, Chicago.
Klee, R. (1997), *Introduction to Philosophy of Science Cutting Nature at its Seams*, Oxford University Press, New York and Oxford.
Kuhn, Th. S. (1962), *The Structure of Scientific Revolutions*, Chicago University Press, Chicago.
Lakatos, I. (1970), 'Falsification and the Methodology of Scientific Research Programmes', pp. 96-196 in I. Lakatos & A. Musgrave (eds.) *Criticism and Growth of Knowledge*, Cambridge University Press, Cambridge.
Suppe, F. (1972), (ed.) *The Structure of Scientific Theories*, University of Illinois Press, Urbana.
Suppes, P (1960), 'A comparison of the meaning and uses of models in mathematics and the empirical sciences', reprinted in P. Suppes (ed.) *Studies in the Methodology and the Philosophy of Science*, Reidel, Dordrecht, 1969.
van Fraassen, B. (1980), *The Scientific Image*, Clarendon Press, Oxford.

9 Realism and Antirealism

The apple tree in the garden is now in blossom. It was pruned last winter, and almost every autumn it bears a rich yield of delicious apples. I never doubt that it is there when my back's turned, as little that I do that the sun continues to exist after it has set. Nor do I doubt that the Prime Minister really inhabits the same world as I, even though I have only ever seen him on television. No one doubts that the world that we see exists when we are not observing it, or merely watching footage of it: the world of sense experience exists, whether or not we are actually in direct contact with it. We are all realists in the sense that we do not believe that we would be able to perceive and make use of the things that surround us, if they did not exist. We can use the car, because it exists – indeed, some would even say that it is because we can use it that it exists.

But what about the world that we cannot perceive – does that exist too? The researcher observes a virus through her microscope; she studies a photo of the atoms in a pentlandite molecule or scrutinizes the tracks of an alpha particle on a monitor. How do we know that these things are real and not simply effects produced by the instruments we use? For we have no direct experience of them through the senses.

Scientific realism is the ontological position which claims that many of the entities postulated by scientists are real even though not apparent to the naked eye. Things do indeed have their own peculiar properties even though we may be unable to observe them, and the theories that purport to explain them are true or false. Conversely, scientific antirealism is the ontological position to the effect that such entities lack the status of real existents and are merely heuristic constructions. Theories about them are, in consequence, neither true nor false.

A claim that runs like a thread through the preceding chapters is that theories are neither true nor false but should be conceived as sets of linguistic rules. It has also been argued that the objects addressed by science are natural or nominal objects. In the present chapter we shall examine the arguments commonly advanced in defence of scientific realism, and so of the claim that theories are true or false. We shall see that the withholding of truth-values from theories does not entail a commitment to denying that unobservable entities exist. A conception of theories as rules for the description of models allows for a form of realism distinct from that expressed in the claim that theories are the bearers of truth-values, for models are used to explain the features of the particular entities that are represented in the model, and such individual explanations are naturally true or false.

We shall begin by setting out the issue of realism as it arises in relation to natural science contexts and then broaden our perspective. We shall discover that the problem receives its own peculiar formulation in the humanities. When consciousness and meaning are the objects of research, ontological commitments change, but the philosophical question of what are the commitments of research, and how such commitments may be justified, remains the same.

A realist position not only entails an ontological commitment. That standpoint also commits those who embrace it to particular stances with respect to how reality relates to our linguistic representation of it, and with respect to whether we are able to offer a sound justification of our beliefs about the external world. Which doxastic commitments attend on this position and the extent of their scope are moot points. We need to distinguish between three kinds of realist commitments:

1. *Ontological realism*: The world exists independent of human minds. Items in the world are as they are, and have the properties they do, irrespective of whether we are able to apprehend them.
2. *Semantic realism*: Sentences about the world have meaning in virtue of their truth conditions, which are independent of whether we are able in principle to establish whether or not they are satisfied. Descriptive sentences are true or false even though it may not be possible for us to show that they are, and in such cases they are said to have verification-transcendent truth conditions.
3. *Epistemological realism*: It is possible for human agents to acquire certain knowledge of the entities that exist in the world, including those to whose existence science testifies.

An adherent of the *strong* realist position will subscribe to all three commitments but a realist might well wish to endorse a *weaker* version of realism, one shorn of the epistemic and/or semantic commitment. A participant in the debate signals an espousal of realism if he or she feels in the least bound by the ontological commitment.

Antirealism represents quite other commitments. Someone espousing this position will deny at least some – but not necessarily all – of the realist's commitments.

1. *Ontological antirealism*: The world does *not* exist independent of human minds (or some other form of consciousness) and their (its) ability to apprehend the world.
2. *Semantic antirealism*: Sentences about the world have meaning in virtue of their assertibility conditions. The truth-value of descriptive sentences is dependent upon whether we are able (in principle) to establish that these conditions are satisfied; that is, upon whether we can verify or in other ways establish their truth-value.
3. *Epistemological antirealism*: It is not possible to acquire certain knowledge of the entities that exist in the world, not even of those to whose existence science testifies.

Here too it is the case that the *strong* antirealist feels bound by all three commitments while a *weak* antirealist will only subscribe to the semantic and/or the epistemic commitment. It is possible, then, to evince realist and antirealist attitudes at one and the same time, and what one calls oneself becomes a matter of definition. Here we define *realism* as the position that entails all three of the realist commitments, whereas *antirealism* will be taken to designate any position which denies one or more of these commitments.

The drama between realists and antirealists has been enacted and re-enacted on the philosophical stage throughout history, and it has not only been the existence of the external world that has been at issue. Equally contentious have been the existence of the past and the future, mathematical objects, universals, dispositions, probabilities, causes, natural laws, possible worlds, moral and aesthetic values, linguistic meaning, and even mind itself. But these themes do not necessarily all figure in the same play: a realist with respect to the existence of the external world is not constrained to adopt that position with respect to moral values or mathematical objects. When the items at issue are those of science, the actors engaged in the dispute are called scientific realists and scientific antirealists, respectively.

Scientific realism

The picture of science drawn by the realist, then, looks like this: In many scientific explanations, reference is made to items and relations that we cannot observe but to which our theories testify. The items and relations that the researcher can see, and the results his apparatus registers, are all the observable results of unobservable causes. Science looks for theories which, by accounting for the facts, are able to offer a true description of reality.

Bas van Fraassen offers the following (well-known) statement of scientific realism: 'Science aims to give us, in its theories, a literally true story of what the world is like; and acceptance of a scientific theory involves the belief that it is true.' Here we encounter all the commitments of the realist in one place: 1. The ontological strand in 'what the world is like'; 2. the semantic strand in 'in its theories, a literally true story'; and 3. the epistemic strand in 'acceptance of a scientific theory involves the belief that it is true'.

But this definition suffers from certain weaknesses. It refers to the aim of science but sets no constraints on what counts as scientific practice. There is no requirement that actual theories be true, and there is no requirement that the researcher have rational grounds or a rational warrant for his or her presumptions regarding the truth of a theory. There is nothing to prevent its being a case of wishful thinking on the part of the realist.

Any characterization of scientific realism must avoid such implausibilities if it is to represent the position in its strongest version. Here I propose to use the label scientific realist to refer to the following claims about scientific theories:

(a) that they can be ascribed a truth-value or an approximate truth-value,
(b) that they seek to approach truth or an approximation thereto,
(c) that their empirical success is evidence of their truth,
(d) that if they are true the unobservable entities they postulate in fact exist, and
(e) that if they are true they will explain observable phenomena.

What is of interest, of course, is whether the realist is able to show that (c) holds, since the other listed features are only persuasive if (c) is true. Obviously, the problem of induction blocks the move from empirical success to the truth of the theory. The realist, however, still counts empirical success as evidence for the truth of the theory. How is that possible?

On this point Putnam has remarked: '[R]ealism ... is the only philosophy that doesn't make the success of science a miracle.' Against this background, the debate about realism and antirealism becomes an empirical one. So the question is: Which of these two positions offers the best explanation of scientific success? The philosopher, in other words, is directed to the same methods as those used by science to justify a particular hypothesis. The method is the inference to the best explanation and the hypothesis is realism. The entire argument may be summed up in the following five points:

1. Science is successful.
2. Two hypotheses offer explanations of this success: (i) realism, and (ii) antirealism.
3. Realism offers the best explanation of the success of science.
4. Therefore, realism is true – or, at least, more likely to be true than antirealism.
5. Therefore, the truth of theories best explains the success of science.

However, this argument is open to criticism. Indeed, some realists have been among those sceptical of this employment of the inference to the best explanation.

The British philosopher, Alexander Bird, correctly points out that scientific realism is not an empirical theory. But if it is not an empirical theory it cannot explain anything at all, he contends, since facts can only be explained by other facts. Realism has no explanatory power. So even though truth is a property of scientific theories, it has no explanatory power in and of itself. What do have explanatory power are the facts corresponding to a true theory. True predictions determine success, but it is the facts that match these predictions that are explained. For Bird, the success of science is bound up with explanation. We use scientific theories in giving explanations, and this is due to the fact that the laws whose truth the theories assume explain the facts derivable from the laws. Since scientists have succeeded in formulating such theories, these laws must exist. For only with such laws presupposed can theories serve explanatory ends.

This explanation of the success of science is problematic, for the explanatory power in question attaches not to the facts, but to our treatment of them. Diverse theories are able to explain the same phenomena and for the realist it can scarcely be the case that they are all true. Few theories have enjoyed as much success as Newton's, and yet the scientific realist has to concede that Einstein's theory explains the same facts on the basis of laws and principles that are different from Newton's. So we must conclude that Bird's criticism of Putnam, and his own attempted alternative realist explanation of scientific success are unsatisfactory. Let us look instead at the explanation of the success of science offered by the antirealist, before turning to assess Putnam's argument.

Scientific antirealism

An alternative philosophical take on scientific theories has it that since they cannot be shown to describe the world as it is, our theories about the world are neither true nor false. This position thus denies that theoretical (unobservable) entities postulated by theories exist; posited concepts of such entities are purely heuristic constructions that serve to combine and systematize large quantities of data. This, then, is the instrumentalist picture of science: scientific theories are intellectual tools for the prediction of observable phenomena. Instrumentalism builds on a phenomenalist epistemology to the effect that objects of knowledge are comprised of our own perceptual experiences, and is only about these that we are able to attain certain knowledge. In other words, the instrumentalist is wedded to ontological, semantic and epistemic antirealism with respect to micro-entities.

Alternatively, the instrumentalist might acknowledge the serious deficiencies of phenomenalism as a theory of experience and hold instead that only non-theoretical entities that we are able to observe are real. The theoretical entities, which we cannot experience directly, are mere hypothetical constructions if the words we use to describe them are presumed to refer to something other than what we can see. All sentences in which allusions to theoretical entities occur acquire their meaning by being reducible to sentences that describe only non-theoretical entities.

As the schema below illustrates, the boundary between theoretical and non-theoretical entities does not coincide with that separating visible from invisible entities:

	Visible	Invisible
Theoretical	Nylon, supernovas	Electrons, neutrinos, etc.
Non-theoretical	Tables, chairs, humans, animals, and plants	Distant stars/large portions of the universe

Classic instrumentalism is founded on a distinction that is not substantiated and for that reason is no longer fashionable among philosophers of science.

Antirealism has undergone radical change and as we shall see there is hardly anyone today – with the possible exception of social constructivists – who in a strict sense adopts the ontological commitment of antirealism. Although calling himself an antirealist, van Fraassen is one in a far weaker sense. His position is a distinctly epistemological one, which he calls *constructive empiricism*. He effectually asserts ontological and semantic realism while remaining committed to epistemic antirealism with respect to unobservable entities.

We can describe the spectrum of scientific antirealist positions in terms of their various stances on the five points that we previously used to identify the scientific realist.

A. The *strong* version denies (a), making (b) and (c) irrelevant.
B. A *weaker* version asserts (a) but denies that (a) implies (d) and (e).
C. *Constructive empiricism* asserts (a), (d) and (e) but denies (b) and (c).

The strong version is represented by phenomenalists; the weaker version by logical positivists in an attempt to reduce theoretical statements to observational ones. But as we saw in chapter 2, this position is untenable.

In his book, *The Scientific Image*, van Fraassen offers a modern defence of antirealism which many philosophers have found challenging. He distinguishes between two forms of scientific antirealism: (i) theories are true if properly interpreted. Interpreted literally, theories cannot be true or false, and (ii) theories must be interpreted literally but do not need to be true to be acceptable. Theories are capable of being truth or falsity. Van Fraassen adopts the latter version as his own; the former he attributes to the logical positivists.

What is the constructive empiricist committed to? According to van Fraassen he is committed to the following conception of scientific theories: 'Science aims to give us theories which are empirically adequate; and acceptance of a theory involves as belief only that it is empirically adequate'. A theory is empirically adequate if it is true with respect to its observational consequences; or, better, if there exists at least one model with which all actual phenomena are consistent. The realist will naturally concede that scientific theories must be empirically adequate. But van Fraassen distances himself from the realist by saying that the belief that the theory is true is irrelevant to the acceptance of a theory. He contends that the acceptance of a theory is identical with a belief in its empirical adequacy in conjunction with an ontological and semantic commitment to what the theory says about the world. Quantum mechanics is empirically adequate and so commits us to a belief in atoms and other atomic objects. The latter is not less important than the former. But, he adds, we can never justify our belief in the existence of atoms, and so we can never invoke that belief to show that theories are true. Consequently, truth plays no part in our acceptance of a theory. All that counts is the theory's empirical adequacy, which is to say its truth with respect to its observable implications.

Bas van Fraassen offers two arguments for constructive empiricism. One is epistemic and the other methodological.

He begins by asserting that we can observe things only with the naked eye, and so it makes an important difference whether we see things with or without the aid of instruments. He points out that there is a difference between observing and observing that. But he also holds that a person who does not know what a tennis ball is, and so fails to recognize that what she is looking at is a tennis ball, still sees the ball with the naked eye. He also concedes that we do indeed see things through a telescope because we could, in principle, travel to the pertinent location and look at the same object in situ at close range. But we cannot see by looking through a microscope or see an electron in the cloud chamber since we would never be able to observe these things with the naked eye.

Van Fraassen holds, then, that we are here faced with a principled distinction between the observable and the unobservable. Ultimately it is science, psychology and physiology which instruct us regarding what we are able to observe and what we are not. As van Fraassen puts it: 'X is observable if there are circumstances which are such that, if X is present to us under those circumstances, then we observe it.' He likewise stresses that observability has nothing to do with existence. There exist without doubt many things that we cannot observe, but precisely for that reason we can have no knowledge of them. Accordingly, our belief in their existence cannot enter into our assessment of a given theory – such assessment is restricted to the truth of the theory with respect to its observable implications, which is to say those observable with the naked eye.

The next argument is a criticism of the inference to the best explanation. It builds on two objections.

The first concerns the fact that this inference may be used with respect to observable entities and phenomena. van Fraassen offers the illustration of the mouse in the wall. If, one night, we hear scratching and squeaking coming from the wainscot and the following day find mouse droppings behind the wainscot, the best explanation is that what we heard in the night was a mouse. We could confirm the hypothesis by getting out of bed to look or by laying a trap and procuring proof positive. But this rule of inference cannot be followed in the case of unobservable things. If we insist on doing so, says van Fraassen, the evidence E that the hypothesis H is true (realism) amounts to no more than mere empirical adequacy (antirealism).

Van Fraassen rejects Gilbert Harman's proposal that hypothesis H is a better explanation than hypothesis H^*, other things equal, on evidence E, if and only if:

(a) $P(H) > P(H^*)$
(b) $P(E/H) > P(E/H^*)$

But this criterion for preferring a hypothesis is not usable here because such probabilities are subjective. They are the expression of the relative strengths of our beliefs with respect to E, H and H^* and should be read as: (a) the belief in H is greater than that in H^*, and (b) the belief in E given H is greater than the belief in E given H^*.

For van Fraassen empirical adequacy is all we can extract from an inference to the best explanation. Alexander Bird has attempted to refute this claim. He holds that the inference to the best explanation can only be an inference to the most true and not the most empirically adequate theory. His argument runs as follows:

1. Explanation consists in facts explaining facts.
2. A theory T indicates the existence of facts K that explain evidence E.
3. If T is the best theory, we are sometimes justified in inferring K and the truth of T.
4. But if T is not true, and K accordingly does not exist, we have no explanation.

5. For *T* to be explanatory, *K* must exist.

But there is no reason why van Fraassen should find the considerations supporting this argument persuasive. First, facts do not explain facts – it is we who make reference to facts in an explanation. Second, explanatory power does not hinge on whether the fact referred to exists. If there is no such fact, the explanation is obviously mistaken, but no less an explanation for all that. A hypothesis may well be the best explanation and still be false – it needs merely to be empirically adequate. Aristotle's answer to the question of why things move was once the best explanation even though it was erroneous.

Van Fraassen's second objection targets the success argument: the realist needs an extra premise to the effect that all regularities in nature call for an explanation. But this requirement leads to the acceptance of inaccessible properties. We know from quantum mechanics that there are regularities that cannot be explained by appeal to a causal mechanism. In 1966 J.D. Bell was able to show that an empirical distinction can be drawn between models with and models without hidden variables. And a variety of experiments show that measurements of spatially separate elementary particles may be dependent on each other, as predicted by models without hidden variables – which is to say, without this connection being explicable by reference to the hidden properties of the particles.

The constructive empiricist also sees theories as empirically underdetermined. Two theories are said to be *empirically equivalent* if they have the same observable implications. For then both theories have a model parts of which fit (are isomorphic with) what we can observe. In such cases both are empirically adequate and we cannot on the basis of experience distinguish between the one and the other theory. But does this mean that we are unable to identify the best explanation?

The realist will object, for instance, that it is not empirical adequacy alone that is determinative of the choice of a theory and the belief in its truth. The inference to the best explanation is made on the basis of such epistemic values as 1. empirical adequacy; 2. fresh predictions; 3. coherence with other theories; 4. theoretical unity; 5. theoretical strength; 6. simplicity; and 7. background knowledge. So empirical equivalence is not the same as evidential equivalence. Two theories can be empirically equivalent and yet evidentially non-equivalent if they have the same observational content but the one exhibits greater coherence with other scientific theories.

There is no need for the constructive empiricist to reject any of this. He need have no difficulty with the claim that there may be other reasons for our accepting one theory rather than another if both prove to be empirically adequate. Nonetheless, he will maintain that these criteria do not imply or in any way warrant belief in their truth (in a correspondence sense.) None of these criteria should induce us to believe that the preferred theory is true rather than false. In antiquity, for instance, people had to decide between Aristarchos' heliocentric world picture and Ptolemaeus' geocentric alternative. Let us assume that these theories were empirically equivalent. However, all the above criteria would favour the Ptolemaic system because it agreed with Aristotle's theory.

An antirealist such as van Fraassen is justified in pointing out that it is not because they lead to truth that criteria have epistemic value, but because they normally give us theories that are empirically adequate. Empirical adequacy cannot be assessed in isolation from new predictions, other theories or general background knowledge. Every observation builds on a manifold of theories and other theoretical preconditions, which feature is congruent with the fact that our acceptance of a theory reposes on the totality of what makes a theory empirically adequate. The application of criteria *qua* epistemic values may be explained as a consequence of a process similar to natural selection. Throughout the evolution of science the satisfaction of such criteria has been shown to be conducive to empirical adequacy.

How is the realist to show that these criteria lead to a true explanation and not merely to an explanation that is empirically adequate? For the realist, the relation between truth and empirical adequacy must be such that truth entails adequacy while the converse does not hold. But we know from logic that the antecedent may be false though the consequent be true. So a theory may well be empirically adequate without being true. The realist is constrained to come up with other arguments for the truth of theories than the inference to the best explanation.

It is not enough, van Fraassen insists, for the realist to stress that that it makes a difference whether water consists of H_2O or XYZ, but that it does not have to make an empirical difference. For if the difference in microstructure cannot be cashed out empirically, the one theory is no more explanatory than the other. He likewise dismisses the idea that it is the task of science to deliver explanations even in the absence of any predictive power attaching to them.

With his criticism of scientific realism and its use of the inference to the best explanation, van Fraassen appears to have cast substantial doubt on the truth of theories or, more accurately, he has put a question mark not simply against our epistemic capacity to determine truth-value, but also against our power to justify the truth or falsity of theories. The realist is simply unable to show that scientific theories have properties that go beyond empirical adequacy and that we need truth in order to be able to decide between theories. But how is the empiricist to explain the success of science which Putnam, for one, thought realism the only philosophical position capable of explaining? As I see it, that question is less than perspicuously put. We need to distinguish between two species of success:

A. Descriptive success
B. Manipulative success

Neither van Fraassen nor the scientific realist makes this distinction. For both, the success of science is bound up with the success of theories. When van Fraassen claims that realism does not explain more about the evolution of scientific theories than does a Darwinist conception, he is right in what concerns A. Only theories that are empirically adequate survive. But is his claim satisfactory with respect to B? Is it the case that reference to empirical adequacy explains how we are able to use entities invisible to the naked eye to generate other entities? That would seem to be far more dubious.

Entity realism

Descriptive success is bound up with scientific theory, manipulative success with the entities of science. Ian Hacking and Nancy Cartwright distinguish between *theory realism* and *entity realism*, and both reject realism with respect to theories while arguing for realism with respect to entities.

The difference between these two forms of realism may be defined as follows:

Theory realism: Theories are literally true or literally false because they describe the world as it is, independent of our minds.

Entity realism: Unobservable items such as molecules, atoms and genes exist, even though the scientific theories about them are not necessarily true or false.

Theory realism has to do with epistemic value; it maintains that science has as its aim the search for true theories, whereas entity realism has to do with facts, that is, with whether *scientific practice* actually involves the manipulation of minute, unobservable things.

Hacking points out that experimental physics offers the strongest evidence for realism. Contact with the physical world occurs through experimental practice, and it is that fact which induces in the physicist the belief that there is a world beyond that which she sees with the naked eye. But Hacking also says that the generality of experimental physicists are realists with respect to entities – not with respect to theories. He distinguishes between experimenting on and manipulating an entity. The mere performance of experiments on a (possible) entity does not commit the physicist to a belief in its existence. If, however, it can be used to perform experiments on other things, a commitment to its existence is concomitant upon that use. Hacking is often quoted as saying of minute entities such as electrons: 'If you can spray them on, then they're real.' We do not, he avers, produce instruments and then infer the existence of electrons in the same way that we test a hypothesis and believe in it because it passes the test. When we construct an apparatus such as, say, an electron microscope, we rely on a small number of rules of thumb about electrons in order to produce other phenomena we want to investigate. In other words, if we can use electrons in an experiment or in the deployment of technical apparatus, we have no option but to recognize such as figuring among the building blocks of nature and technology.

We are not, then, committed to a belief in things with which we can have no physical or causal contact. The argument for entity realism can be formulated more precisely as follows:

For any supposed entity X, if it is possible successfully to manipulate X, then X exists.

We do not believe in elves or gremlins because no such creatures have ever bothered us and nor do we believe in the existence of unicorns – having never ridden on such or eaten their meat. This makes it possible for us to explain why we are committed to the existence of X. Practical success warrants belief in referential success. Descriptive success, by contrast, offers no guarantee of such success. We can depict elves and unicorns to our hearts' content without their needing to exist.

The theory realist attacks the entity realist with the criticism we would expect: To manipulate a theoretical entity successfully, one must have some knowledge of its causal properties, but having such knowledge implies the theory must be true. The conclusion follows that entity realism entails theory realism. It is, however, difficult to take this objection very seriously. We do not need a theory of elves, gremlins or unicorns to be able to come into physical contact with them. Such beings must obviously have properties for us to be able to do anything with them at all, and we must know something of these properties in order to adapt our behaviour accordingly. For example, one cannot eat unicorns without knowing whether they are edible. But such knowledge does not amount to a theory of unicorns. By the same token, it is possible to build an electron microscope without reference to the Schrödinger equation. Many of the properties of the electron were discovered long before the theory of the electron, which first emerged with quantum mechanics. And all that is required of a candidate theory is that it be empirically adequate.

Nancy Cartwright also subscribes to entity realism but her argument is a somewhat different from Hacking's. On her view, belief in the existence of a thing is warranted if we can experiment on it. We do not need to have progressed to the point at which we can use the item to manipulate other things. Experiments supply us with the basis *on which to infer the most probable cause*. She argues that inference to the best explanation is not a valid argument for the existence of particular entities. As she says: 'No inference to the best explanation, only the inference to the most probable cause.' She rejects the inference to the best explanation on the grounds that realists use it to assess the truth-value of theories and not merely in order to determine whether entities exist.

The scientific realist uses the inference to the best explanation to assert the (approximate) truth of a given theory.

Inference to the best explanation

1. P explains Q.
2. Q is true.
3. Therefore, P is true.

This inference is acceptable, says Cartwright, only if *P* indicates a cause, rather than a fundamental law. We can use causes to explain, but not all explanations are causal explanations.

Inference to the most probable cause

1. *P* causes *Q*.
2. *Q* exists.
3. Therefore, *P* exists.

This inference offers good grounds for a belief in theoretical entities.

To illustrate how an inference to the most probable cause may be established, Cartwright adduces Perrin's justification in 1913 of the existence of atoms and molecules – an argument offered at a time when their reality could still be reasonably doubted by contemporary instrumentalists. Jean Baptiste Perrin showed that the Avogadro constant, which specifies the number of molecules in a particular weight of matter, could be determined on the basis of thirteen different and mutually independent types of experiment. Cartwright's point is that the belief that molecules are the cause of the concordant results has a far greater probability than any alternative hypothesis, and that that probability is quite independent of any more exact elaboration of atomic theory that, over time, has received various formulations. There is a causal story to be told and it is either true or false.

Cartwright's argument is based on Reichenbach's principle of the common cause. If two events, *A* and *B* are correlated, but without *A* being the cause of *B* nor *B* the cause of *A*, there must be a common cause, *C*. The argument is as follows:

1. A positive correlation between events *A* and *B* requires a common cause.
2. Among statistically positively correlated events there is often no observable common cause.
3. Therefore, there must be a common cause *C*, which is unobservable.

We have, then:

(a) The probability of *A* given *B* is defined as $P(A/B) = P(A\&B)/P(B)$

where *A* and *B* are two events, and $P(B) \neq 0$.

(b) *B is statistically relevant* for A, iff $P(A/B) \neq P(A)$.
(c) There is a *positive correlation* between A and B iff $P(A/B) > P(A).P(B)$.
(d) From (a) and (c) it follows that if $P(A) \neq 0$ and $P(B) \neq 0$ then A and B are positively correlated, iff $P(A/B) > P(A)$, and also iff $P(B/A) > P(B)$.

We can now give a precise formulation of the common cause as:

$P(A/B\&C) = P(A/C)$.

In other words A and B are statistically mutually independent relative to C.

We can accordingly define a relative probability in relation to C, conditionally or unconditionally, respectively:

(e) $P_c(X) = P(X/C)$,
(f) $P_c(X/Y) = P_c(X\&Y)/P_c(Y) = P(X/Y\&C) = P(X/C)$,

where $P_c(Y) \neq 0$ and $P(C) \neq 0$.

In other words we can say that C is the *common* cause of the correlation between A and B if, relative to C, there is no such correlation.

What does van Fraassen say to this? On this point as well he stands as the exponent of a latter-day antirealism even with respect to entities. There are, on his view, no epistemic routes to the validation of the belief that there exist unobservable entities such as atoms and molecules. We saw earlier that in quantum mechanics positive correlations may be encountered which only have a common cause given the postulation of hidden variables. So van Fraassen contends that Reichenbach's principle has only limited validity, even in the case of observable entities, and we cannot extend its applicability to items that cannot be seen with the naked eye.

In my view, there are grounds for questioning van Fraassen's entire argument. He is of course right in claiming that two independent atomic events, A and B, do not need to have a common cause, even though statistically correlated. In an indeterministic world such phenomena will occur, and quantum mechanics suggests that we live in precisely such a world. But he is mistaken if, putting atomic events to one side, we let A and B stand for the sortal properties of things. His claim then rests on a spurious distinction between what can and cannot be observed with the naked eye. For it is impossible to specify pertinent epistemic differences between perceiving with the naked eye and observing with the aid of instruments. We see visible things but we can observe invisible things. In both cases we see a thing as an item of a particular kind because we can apply a name to it. And we can apply a name because we are familiar with the criteria which license the use of the name and are able to recognize circumstances that meet them. Such evidentiary criteria enter into the meaning of the name, offering us an assurance that the name normally refers to what it is that the criteria pick out. Noteworthy among such criteria are visual and operational evidence for the use of the name. If we are able to manipulate an electron, some of the criteria for the application of that term are fulfilled and we are logically committed to the belief that electrons exist. Strictly speaking, the criteria will very often be fulfilled if we are simply able to experiment with the item in question. But we are not, *pace* Cartwright, obliged to believe in electrons because they are the most probable cause. We are committed to a belief in unobservable entities if it is part of their definition that the available evidence is good evidence for their existence. It is a slight but crucial difference. The commitment Cartwright argues for is contingent, the one I propose, necessary.

We can conclude, then, that van Fraassen has succeeded in showing that the scientific realist cannot use the inference to the best explanation as an argument for the truth of theories, but he is much less successful in undermining our commitment to the belief that there are unobservable things, and that we can show that they exist.

Social constructivism

Science is constituent in a larger social activity that evolves through the cooperative efforts of many people. Researchers, technicians, administrators and politicians contribute to the articulation of goals and to the advancement of research in directions beneficial to society. Patently, then, the evolution of science, as a function of historical, epistemic and social vectors, can itself be the object of empirical studies. Science studies is the meta-level discipline that concerns itself with such research. It comprises the history of science, along with that of psychology and sociology. These subjects are no different from others; they must be assessed on the basis of the same methodological requirements as those that apply to any other empirical science.

However, a scientific orientation has appeared on the scene in recent decades that seeks to focus on knowledge – including scientific knowledge – as a 'socially constructed object' and hence to be explained sociologically rather than through the philosophical analysis of epistemic justification. This discipline is called the sociology of knowledge and builds on an empirical foundation. But the sociology of knowledge is often elevated to the status of a philosophical theory, and as such has been given the name 'social constructivism'. Whereas traditional theory of knowledge is essentialist – it sees knowledge as an objective entity, its content determined by its object – social constructivism is anti-essentialist, in the sense that it holds knowledge to be determined by social norms and rules.

A distinction is made between the weak and strong programmes of social constructivism. The *weak programme* is of no interest in the present context. It simply holds that social factors play an important role when new theories are formulated. Few wish to deny this. The *strong programme*, by contrast, wants to topple the philosopher from his Olympian heights. Athena does not understand art and nature because she sprang from the forehead of Zeus. She understands the world because there is collective agreement among the gods as to what she should understand. Scientific knowledge must be understood in the light of *nomos*, not of *logos*. But in their zeal to expel the philosopher from Olympus this theory's proponents explode the mountain itself and with it the dwelling-place of the gods. They ought, rather, to take the trouble to clamber up over the clouds where they would find that Athena keeps the lightning in her chamber. For just as *logos* may not be drawn on to explain why mortals create the scientific theories that they do, *nomos* is no more fit to explain the practical knowledge those same mortals possess.

David Bloor is the champion of the strong programme. It is quite unacceptable, he contends, that non-rational social causes are used to explain the errors of science, while scientific success is explained on the basis of its congruence with nature. The success of science cannot be explained by reference to the truth of our beliefs. On the contrary, its success is the consequence of a social accommodation of what we believe to be true with what society generally holds to be true. The basic thesis is that *scientific knowledge is socially constructed and not discovered*. Bloor denies that this thesis entails idealism (ontological antirealism) and relativism (epistemic antirealism). For him, social reality has the same objective status as that enjoyed by nature for the traditional philosopher. The core idea is that the material world is filtered through *socially sanctioned metaphors*, so that its nature and structure play no significant role in determining the *content* of our scientific theories.

What, then, is Bloor's argument? His contention is that true beliefs in science are merely the result of data being interpreted via the matrix of social metaphors, just as are false beliefs. So the content of scientific beliefs are socially determined in the same measure as are the contents of all other beliefs, and this means that they are more socially determined than not. Consequently, scientific knowledge is socially constructed rather than determined by the character of a given world.

One question is whether Bloor is correct in asserting that this conclusion does not entail idealism and relativism. The conclusion does not follow from the premises. Let us concede that it holds of true, as well as of false beliefs, that they are each the result of a socially sanctioned interpretation of particular data. There is nothing surprising in this inasmuch as all scientific knowledge is expressed in language, and the use of language is as least sometimes accompanied by interpretation. But how have the data of science themselves been obtained? If they are the products of an interpretation there must be something else that mediates that interpretation, which is not in itself an interpretation. There must be objective facts that are not socially caused but are that from which an interpretation takes off, and which we understand *qua* uninterpreted. They are the facts reported by our senses and which our beliefs are beliefs about. Only on this construction is idealism or relativism avoided.

In fact, however, it never becomes clear whether social constructivism is an ontological thesis as well as an epistemic thesis. Our knowledge of the world may well, in part, be constructed without the world so being. But then it makes no sense to say that we can never have objective knowledge of this world. For how else would we know that the world is not constructed?

The next step for the social constructivist is that of explaining why certain beliefs are successful when this success is not bound up with their articulating how the world is. Here Bloor points to its practical success.

1. Why does science firmly adhere to the belief B?
2. Because the acceptance of B yields practical success.

The entity realist too would be able to point to 2 as the answer to 1. But to the following question:

3. Why does the acceptance of *B* yield practical success?

he will answer:

4. Because *B* is true (of a theory-independent reality).

The social constructivist cannot respond to 3 by offering 4 but has to say:

4.* Because prevailing social metaphors license talk of success when the belief *B* is sustainable in a social practice.

The alleged success is thus generated by special social metaphors and connections that are projected onto nature via scientific methods and theories. There are no objective limits to the extent to which social metaphors and connections can come to dominate a culture. In principle this means that, in science, any belief can potentially be 'socially justified.'

If, then, the strong programme is taken literally, the term 'success' is just as socially determined a metaphor as 'reality' and 'truth'. There is no material practice in which such success is founded – only a social practice.

Bloor's counterargument consists in his insistence that the social is the objectively real world and that there are, in consequence, limits to the extent to which social metaphors can be used to explain social constructivism. For if the social constructivist's beliefs are themselves no more than socially sanctioned metaphors – why should we bother with them? In this way Bloor seeks to stave off the very same criticism which he uses to attack the realist.

Bloor's argument does not stand up. For either the data are constructed, or else they are objectively present. If they are the expression of an interpretation because generated by an interpretation of social practices, then social constructivism rests on one out of many interpretations of the social data. And if these data are objectively present, this only ostensibly explains the success of social constructivism. For since there is no argument to show that social data are to a lesser degree constructed and so more objective than those of natural science (most people would think that the converse held), social constructivism can hardly be the key to understanding the track record of science.

In Bruno Latour and Steve Woolgar we meet a line of thought that in many respects reminds us of Bloor's. In their book, *Laboratory Life: The Construction of Scientific Facts,* they advance a number of striking claims based on field studies conducted in the laboratory. The scientific community must, they contend, be studied in the same way as are 'primitive' cultures by the anthropologist. The self-understanding of the culture should not be taken as an accurate account of regnant social practices. Should a hypothetical entity or fact over time acquire the status of a real entity or fact, owing to increased confidence in that thing or fact on a scientific perspective, then that item is a result of the social act of creation. An external 'object' becomes a 'concretized' *invention* because of the repeated attention given a particular 'statement' in journals and textbooks. The objectivity is only apparent. At one point they say: 'Despite the fact that our scientists [*i.e.* those Latour and Woolgar studied in the laboratory] held the belief that the inscription could be representations or indicators of some entity with an independent existence 'out there,' we have argued that such entities were constituted solely through the use of these inscriptions. It is not simply that differences between curves [on superimposed graphs] indicate the presence of a substance; rather the substance is identical with the perceived differences between the curves.' (p. 128) The quotation well illustrates the deficiencies of Latour and Woolgar's analysis of experimental practice. It is impossible – unless one is prepared to embrace out-and-out behaviourism – simply to ignore researchers' own understanding of what they are doing. In many cases the signs commit the researcher to believing that what the signs stand for actually exists because reference to them is constituent in the definition of the object in question. Latour and Woolgar's standpoint is ultimately a latter-day version of instrumentalism, with the logical construction of theoretical entities of an earlier era being replaced by a social construction.

Social constructivism's challenge to scientific realism does not touch the development the philosophy of science has undergone over the past two decades which this book has recounted. The evidence itself (the signs) figures in the criteria that lead us to believe that there is a world 'out there'. But our commitment to the belief that there are unobservable entities does not commit us to believing that scientific theories concerning these entities are true or false. Rather than expressing factual beliefs, they furnish us with the vocabulary and rules for talking about reality. They are indeed more the expression of *nomos* than of *logos*. It is only in scientific practice that *logos* dovetails with *nomos* in our encounter with the world.

The human sciences

Within the social and human sciences as well we find a debate between realists and antirealists about the nature of their goals. No one doubts that concrete entities such as society, people, books and paintings exist. This is not the problem. The problem arises at the point at which we have to decide whether unobservable or theoretical entities such as inflation, intentions, and/or unconscious mental processes exist, and whether they play an active role in our understanding of society, people, books and paintings. What is interesting is not whether Marxism, structuralism or psychoanalysis is literally true or false – for they are neither. The important question here is whether they prove to be empirically adequate when drawn on to describe economic processes, texts or the human psyche. Whether they are empirically adequate is not a question for philosophy but is one to be settled alone by those who practise the science in which the theory receives application.

Just as in the discussion about the natural sciences, we need to address the position that what the models are about – the entities to which they relate – exist, even though not observable with the naked eye, and that such entities are significant when we seek to understand the properties of the objects and processes that we see, read, hear, feel, etc. But so as not to lose sight of our aim, let us once more turn the spotlight on the interpretive sciences.

Discussions about correct interpretation often refer to what is called the intentional fallacy. This expression is due to W.K. Wimsatt and Monroe Beardsley who, in a jointly-authored article, argued that a confusion of the question of the work's meaning, understood as its *intension*, with the author's or artist's meaning understood as that individual's *intention*, should be denominated a *fallacy*. Against those critics, I shall argue that there is no fallacy here but that the distinction between the meaning intended by the author or artist and the meaning of the work gives rise to a discussion about realism and antirealism within the interpretive sciences paralleling that conducted within the natural sciences which we have considered in the previous section. Any differences there may be derive purely from the greater complexity that here attaches to the problem.

We need to isolate four distinctive positions in the debate which we will call:

A. Intentionalism (realism)
B. Anti-intentionalism (ontological antirealism)

C. Interpretive instrumentalism (epistemological antirealism)
D. Hypothetical intentionalism (entity realism)

The author's intentions are the unobservable factors that mediate the work's meaning. The intentionalist is a realist. He holds that the interpretation of a literary text is a theory of authorial intention, and that such theories are determined as true or false by reference to such intentions. The structure and meaning of the work is the evidence we have for saying that an interpretation is true in virtue of its successfully capturing the author's intentions, and that we explain the meaning of the work by reference to such intentions.

The anti-intentionalist, by contrast, who holds that the work is semantically autonomous, embraces a strong antirealist position. A position is not true or false by corresponding or failing to correspond to the author's intentions. The notion of authorial intention is merely a heuristic construction.

We need also to distinguish between interpretive instrumentalism and hypothetical intentionalism. The interpretive instrumentalist sees the textual meaning of the work as that which makes it possible for us to frame a hypothesis about the author's intentions, though not a hypothesis that offers an objective representation of them. Against this view, the hypothetical intentionalist asserts that it is precisely an objective representation that the hypothesis delivers. The work serves as just that empirical basis on which the interpreter can construct an objective understanding of the author's intentions.

Anti-intentionalism was essentially the foundation for the New Criticism in the 1940s, 50s and 60s, but already a figure such as Heidegger had adopted a similar stance. More recently it has become the foundation of the infamous dictum of postmodernism: 'The only author is a dead author'. In modern times E.D. Hirsch has defended intentionalism, and later Gregory Currie has argued for interpretive instrumentalism, while William Tolhurst and Jerrold Levinson have sponsored hypothetical intentionalism. Briefly, then, these represent the chief positions in the debate.

But what is it, exactly, that we seek to understand? It is, indeed, the meaning of the work. But what does a literary text 'mean', and in what does this meaning consist? We shall consider four proposals:

1. Words' *sequential meaning* (literal meaning).

2. *Utterer's meaning* (that intended by the author through his or her choice of words).
3. *Utterance meaning* (the meaning which such sentences receive through being presented in a context in which the sentences are articulated by that particular author).
4. *Metaphysical meaning* (the meaning that arises through the interplay of various interpretations in the light of plausibility, intelligibility or interest).

The first proposal, which refers to the *literal* meaning, is an option only if we refrain from distinguishing between literary texts and all other linguistic utterances. But a literary text is distinctive: it differs from, say, a booklet about walks in the Mendips published by the Ramblers' Association by not being written anonymously. We want a literary text to have a particular intention through being the product of an individual mind. So option 1. designates the meaning of the text but not its literary meaning.

This might lead us to think that we should turn, instead, to the intention of the originator of the text. But the author's intentions can never be synonymous with literary meaning. Utterer's meaning is a mental state possessed by the author, and the literary meaning of the work is logically distinct from the author's aspirations in writing his work.

Nor is the last proposal satisfactory. For metaphysical meaning presupposes the presence of a more fundamental kind of meaning.

The meaning of the utterance seems to be what we might call the literary meaning. It does not rest solely on the meaning of the text but also on the context in which the author chose to say what she did. Tolhurst defines utterance meaning as the intention that a member of the intended readership would be most justified in attributing to the author on the basis of the knowledge and attitudes she possesses *qua* member of that readership. Utterance meaning is to be regarded as a *hypothesis* about the utterer's meaning that an individual, *qua* intended reader or hearer, might appropriately proffer in the light of her background knowledge and attitudes. However, in the present context it is more relevant to speak of a researcher rather than the typical, intended reader. The researcher is distinguished from the ordinary reader by being committed to understanding the work *qua* utterance – the ordinary reader may simply read the work for pleasure, she does not need to justify her particular reading of it.

Levinson adopts Tolhurst's term 'hypothetical intentionalism' but cannot endorse his definition. They diverge insofar as Levinson operates with the notion of an ideal interpreter, drawing not on the author's actual intentions in relation to his readership as does Tolhurst, but seeking instead to reproduce the appreciation one would find in an appropriate or ideal readership whose members are apparently also omniscient. This ideal interpreter approximates more closely to the role of scholar, even though the latter is by no means omniscient. Levinson holds that it is the task of the ideal interpreter to grasp the author's *categorial* intentions, *i.e.* those expressed in virtue of the species of work the author sought to create. Insofar as the work's text and structure is in line with these intentions, the ideal interpreter must include them in his account of the work's literary meaning. But in what concerns the author's *semantic* intentions, the ideal interpreter may depart from the author's actual aims, so long as the interpretation is compatible with the text, the author's categorial aims and the various contextual features. Levinson holds that the difference between literary texts and other texts is that the meaning of a literary text, as understood in terms of utterances offered by both author and reader, is determined by the text's inherent potential. He is contending that the meaning of the utterance must always be understood in light of a context comprised of author, readership, period, cultural background, predecessors, history of the genre, contemporary events and the author's oeuvre as a whole. This, according to Levinson, makes it possible for a reading to raise the artistic value of the work beyond what the author's own semantic intentions would permit.

What Tolhurst and Levinson have in common, however, is not merely the name 'hypothetical intentionalism', but that they each contend that the author's actual intentions have determinate properties in virtue of their specific content, that the work expresses that content, and that claims about such intentions must therefore be true or false independent of any theory of interpretation.

So we need to distinguish between *textual* and *literary* meaning and to explain how we are to understand such meaning. On the one hand, we have the problem of the author's intentions in contrast to the work's *literary* meaning: whether the work's literary meaning functions as a vehicle for what is, for the reader, an inaccessible reality in the form of the author's intentions. On the other hand, we have the issue of the work's *textual* meaning in contrast to its literary meaning: the textual meaning reveals itself in the form of the semantic content of the sentences, while their literary significance exists in virtue of the allegorical, metaphorical and contextual use of the language not manifest in the literal reading.

An antirealist with respect to *textual* meaning would assert that there is no invariant or literal meaning, and that, consequently, the textual content shifts from one language user to another. This is the position represented by the literary scholar Stanley Fish. His realist counterpart would defend the constancy of textual meaning on the grounds that there exists such a thing as literal meaning, communicable by one person to another, and which mediates the metaphysical and symbolic use of language. Indeed, it would be in line with the realist's fundamental position here to assert that such invariant and literal meaning is a precondition of our being able to use the language allegorically and metaphorically at all.

Being an antirealist with respect to *literary* meaning does not commit one to a comparable stance with respect to textual meaning. But the implication holds in the other direction: if one is an antirealist with respect to textual meaning, one must also be one with respect to literary meaning. One might perhaps, stretching concepts a bit, compare the relation between the author's intentions, literary meaning and the textual meaning with that obtaining between the micro-world, macro-world and our phenomenal experience of this macro-world. Here there is room for two different types of ontological antirealism: micro-phenomenalism and macro-phenomenalism. An antirealist stance with respect to micro-entities does not commit one to adopting the same stance in relation to macro-entities, but if one takes up that position to the macro-world then, necessarily, one also does so with respect to the micro-world.

In the present context we shall concentrate on elucidating the philosophical problems attaching to literary meaning. To what extent can we explain the work's textual meaning as mediator of the author's intentions, and to what extent can the interpreter use such meaning to formulate a hypothesis regarding its literary meaning, which may in turn shed light on the author's intention?

A scientific realist with respect to *literary* meaning holds that the various theories of interpretation may be used to tell us something about the author's intentions, and it is these that interpretative science has to identify in order to explain the work as a whole. Such theories are true and it is the aim of literary studies to show that these interpretations offer a literally true account of what the author's intentions were, and to accept them is to recognize their truth relative to the author's objectively real intentions.

The *strong* ontological antirealist will claim that the text is semantically autonomous; a correct understanding of the text is wholly independent of the author's purposes in writing it. Literary meaning does not extend beyond the text's publicly accessible meaning. Indeed, if he is also a strong ontological antirealist concerning textual meaning he will claim that literary meaning reduces to the understanding of the individual reader. Every reference to authorial intention is merely a heuristic construction, just as literary theories are simply tools with which to ascertain the structure in the text. Anti-intentionalism and instrumentalism are one and the same.

But it is every bit possible for someone to be an ontological realist but a semantic and an epistemological antirealist or, as in the case of Bas van Fraassen, an ontological and semantic realist but an epistemological antirealist. The *weak epistemological* antirealist holds that since literary theories are often *intensionally underdetermined*, they can say nothing about the author's intentions, and so the task of the interpretive sciences is simply to deliver *intensionally adequate* theories. Gregory Currie articulates just such a line of thought. He explicitly refers to Bas van Fraassen as an influence. He sees the antirealist as one who seeks to construct a hypothesis about the author's narrative intentions, but without the hypothesis necessarily being assumed to be true, as it is for the realist. What is crucial to the acceptance of an interpretation is not whether the hypothesis actually matches the author's literary aspirations, but whether it is *intensionally adequate* in relation to the text.

When Currie refutes the notion of the objectivity of authorial intention, this rests on his understanding of the best interpretation. For us to be able to speak of the best interpretation there needs to be consensus across the spectrum of interpretations regarding their relative merits. Even when two interpretations are deemed to be of equal merit, that judgement must proceed on the basis of common criteria. But there are no such criteria according to Currie. And even were a tolerance constraint to specify that from the perspective of a determinate set of criteria (not necessarily endorsed by another interpreter) there is in fact a 'best explanation', Currie doubts that any such constraint could be set up. He holds, that is, that we require a measure of value-relevance in an interpretation before finding it acceptable, and that we always connect the plausibility of an explanation with its value for us in the light of our background and system of values.

There is, then, reason to distinguish between a stronger ontological and a weaker epistemological version of interpretive antirealism: the *ontological* version assumes the absolute literary autonomy of the work, inclusive of its metaphorical and symbolic meaning, while the *epistemological* version accepts that a hypothetical set of intentions is indispensable because the symbolic or implicit content can only be construed if the text is *intended* to mean something over and above what is expressed in terms of its literal or explicit meaning. The epistemological antirealist does not doubt that there are (or were) such intentions, he merely asserts that in principle we will never have the means whereby to ascertain whether the hypothesized intentions are true or false, because we never share the same values and criteria.

An important difference between the natural sciences and the interpretive sciences stands out. Inanimate nature forms part of the subject matter of the natural sciences and it is not possible to gain direct perceptual access to the micro-world. All speculation and theorizing about atoms and molecules must rest on our observations of the macro-world. This does not necessarily apply to authorial intentions. Here the artist, if living, might – directly and not through the work – inform others of what were his or her intentions in writing the text. But there is more. The author might also let it be known that a particular interpretation fails to tally with his or her intentions. And lastly, there might be a conflict between the interpretation the author is prepared to accept and that which the interpreter produces.

Precisely because the author is able to speak of his or her intentions, literary studies has, in that respect, an edge on natural science. In literary studies the researcher may sometimes be afforded direct access to what, in the normal course of things, is accessible only through the text itself. He or she may sometimes receive direct confirmation that there are narrative intentions embedded in the text to be uncovered. A realist will find it wholly legitimate to draw on the author's own perspectives when framing an interpretation of the work since he holds that literary theories about its meaning are true or false depending on whether they correspond to the author's actual intentions. Indeed, in defending his position he will adduce the fact that whenever we are allowed access to them, there prove to be such intentions; and for that reason it is the role of the interpreter to uncover them.

To the extent that a report from the author contributes to the interpretation of the work or proves to coincide with the interpreter's reading of the text it poses no problems for the realist. A problem arises only when a given interpretation does not figure among the understandings of which the author was aware, or is in direct conflict with those that he or she was. This difficulty is one the antirealist is swift to exploit.

When it is a matter of the author's intentions in relation to the meaning of the literary work, there would seem to be four possibilities:

1. The author's conscious intentions are expressed by the *literary* meaning of the work.
2. The conscious intentions do *not* come to expression in the *literary* meaning of the work.
3. The work's *literary* meaning expresses intentions that the author did not have (or was not aware of) but which the author is prepared to accept as being consistent with the *textual* meaning of the work.
4. The *literary* meaning expresses intentions that the author utterly disavows and which he or she might also maintain cannot be a correct interpretation given the work's *textual* meaning.

The antirealist then argues that it is possible that the intentions of the author on the one hand are not satisfactorily expressed in the literary meaning of the work or that the text may indeed possess literary meaning not intended by the author. There are examples of an author's being dissatisfied with her finished work because it fails to express the emotional timbre at which she had aimed, or to fulfil the literary ambitions that she had had in her sights. We are also well aware that the mainsprings of our actions are often obscure or hidden from us.

The strong antirealist is in no doubt: 2 and 4 hold. There can be no relevant causal relation between authorial intention and textual meaning, and so an interpretation of the textual meaning, *i.e.* the literary meaning, can tell us nothing about the author's intentions. He thus concludes that the text possesses absolute semantic autonomy. By contrast, the weak antirealist will hold that we do know that the work has been produced with certain ends in view. A theory of interpretation must construe the author's intentions as the causes of the textual meaning. However, we can never validate the truth or falsity of our beliefs. We can only determine whether they are intentionally adequate.

The realist finds herself engulfed in problems. She will probably accept 3 along with 1. But with regard to the muster of arguments for the claim that theories of interpretation give us a true picture of the author's intentions she remains deficient. The reason for this is simple: such theories are not concerned with specific intentions – what they furnish us with are the linguistic resources to interpret concrete works. If the realist is to meet the challenges of the antirealist, he must qualify his position so that it becomes a species of entity realism. The textual meaning of the work represents the author's intentions and the interpretation of this representation is its literary meaning.

Both the theory realist and the entity realist hold that the author's intentions are real entities that have an explanatory function to play in a correct understanding of the literary meaning of the work. According to both, the interpretation of a text explicitates the author's intentions. But whereas the entity realist sees literary meaning as an interpretive hypothesis about author's intentions in relation to the text, the theory realist sees the theories of interpretation as true of the literary meaning if they represent the author's actual intentions: which is to say that an interpretation purports to reflect what the author herself might say about her text. The entity realist rejects the claim that theories of interpretation are committed to serving as literally true descriptions of the content of the author's intentions. She believes that the author's actual intentions have certain properties, namely the possession of this or that categorial content, and that claims about these intentions can be true or false independent of any theory of interpretation. Thus, structuralism is not a true theory about the author's literary intentions as they are expressed in the text, but its principles may be used to formulate a model of the author's intention in order to find out what she might want the concrete text to say. Just as, for instance, neither Marxism nor psychoanalysis is true or false, nor are theories of interpretation. Such theories do not give us a literally true or false description of the literary meaning in terms of the author's intentions, but they may, when used to describe a model of the author's intentions – a model which has been abstracted from the textual meaning of the work – assist the researcher in the construction of a literary interpretation of the text. What are true or false are the concrete explanations of the text's literary meaning, the interpretation showing how the text expresses the author's intentions; by framing a model which is already implicitly present in the text *qua* formal structure, by offering specifications in terms of genre, sub-genre etc., the researcher is able to say something about the intentions informing the working out of the text.

To begin with it is clear that it is prerequisite to the meaningfulness of a text that it be produced by an author with the aim of conveying meaning through it. Were there no author, there would be no text to understand, and without the authorial aspiration to communicate in literary mode, there would be no literary value.

It is not incorrect to say that there may well be cases where a correct interpretation is not evident to the author herself. Here Levinson's distinction between the author's categorial and semantic intentions proves its worth. The author's semantic intentions may be, in part, unconscious but would, however, be acceptable to the author were she presented with them, as congruent with the work's semantic content. Here we can say that we are making *an inference to the most probable* intention against the background of the work's textual meaning. Finally, we may find ourselves in a situation in which there are various interpretations on offer, and neither the author nor the interpreter can determine which of them best captures the semantic intentions expressed by the text. In such cases the antirealist is right to hold that such intentions are *intensionally underdetermined*. But it cannot be inferred from this that no intentions are present, or that those that are, are irrelevant to the understanding of the work. So the antirealist has not made her case.

The entity realist has not forfeited his right to concede that our interpretations may be intensionally underdetermined. This is because intensional underdetermination does not concern the ontological domain but rather the limitations of method in relation to our epistemic goal of truth. This epistemic goal will be one the realist can retain notwithstanding the impossibility of making a determined choice from among a diversity of interpretations against the background of textual meaning.

Second, the antirealist skates over point 4. This point raises the issue of over-interpretation. If the interpreter has exclusive regard for the meaning of the text and disregards what the author intended with his or her text, then the door to relativism is flung wide open. Irrespective of whether one is a strong or weak antirealist, problems arise from any conflation of the reader's values with those of the author. Such a conflation is consistent with any and every interpretation since in principle every text might speak differently to different readers. We avoid relativism by distinguishing between the interpretation of the ordinary reader and the scholar's interpretation of the work. Indeed, even with that distinction in place it is not enough to say that epistemic aims such as intensional adequacy, explanatory simplicity and coherence, circumscribe possible interpretations. They still permit too great a variety. Only by including the author's intention as the cause of the textual meaning is the researcher able to elaborate a sufficiently well-defined target for candidate interpretations to meet, if the scope of possibly valid interpretations is to be appropriately circumscribed. Naturally, this does not lead to the interpreter being able to decide whether an interpretation is an over-interpretation only when it is possible to refer directly to the author's intentions. Often she must settle for much less, but it must still be the case that none of the criteria set up for the evaluation of over-interpretation gives results at variance with those which would emerge on the hypothesis that we could ask the author what her true intentions were.

Hypothetical intentionalism sees intentions, literary meaning and textual meaning as logically discrete entities that are causally and conceptually linked. The author's literary intentions and her work exist in the world, and literary meaning is an interpretation of the connection between them. The writing of literature is a communicative act that takes place in a particular context. Textual meaning is the goal of that act and literary meaning is a hypothesis as to why the act took the form that it did. As we saw in the discussion of entity realism, we are committed to a belief in unobservable entities if it is part of their definition that the evidence currently available to us is good evidence for their existence. And as we have argued above, the relation between motive and action cannot be analytic, since it is part of our concept of motive that the act that springs from a given motive constitutes good evidence for the motive itself. In sum, these considerations indicate that textual and contextual meaning work as sortal criteria of the author's literary intentions, and that our concept of authorial intention is such that textual and contextual meaning offer good evidence of those intentions even though the evidence may prove fallible.

Literary meaning must not be confused with what we here have called metaphysical meaning or what Levinson calls ludic meaning. I prefer the former term. It is not often that we encounter a distinction between the literary and the metaphysical meaning of a work. But in the human sciences, as in the natural sciences, interpretations are advanced that go far beyond what can be established by relying solely on empirical resources, simply because our hypotheses are empirically or intensionally underdetermined. In our final chapter we shall take a closer look at this issue.

References

Bird, A. (1998), *Philosophy of Science*, UCL Press, London.
Bloor, D. (1991), *Knowledge and Social Imagery*, Chicago University Press, Chicago.
Cartwright, N. (1983), *How the Laws of Physics Lie*, Clarendon Press, Oxford.
Currie G. (1993), 'Interpretation and Objectivity', in *Mind*, 102, pp. 413-428.
Faye, J. (2000), 'Observing the Unobservable?', in E. Agazzi & M. Pauri (eds.) *Observability, Unobservability and the Issue of Scientific Realism*, Kluwer Academic Publishers, Dordrecht.
Fish S. (1980), 'Is There a Text in This Class?', in *Is There a Text in This Class? The Authority of Interpretive Communities*, Harvard University Press, Cambridge Mass. pp. 303-321.

Hacking, I. (1983), *Representing and Intervening*, Cambridge University Press, Cambridge.
Hirsch, E.D. (1967), *Validity in Interpretation*, Yale University Press, New Haven.
Klee, R. (1997), *Introduction to the Philosophy of Science. Cutting Nature at its Seams*, Oxford University Press, New York and Oxford.
Latour, B. and Woolgar, S. (1986), *Laboratory Life: The Construction of Scientific Facts*, Princeton University Press, Princeton.
Levinson, J. (1996), 'Intention and Interpretation in Literature', in *The Pleasures of Aesthetics*, Cornell University Press, Ithaca.
Salmon, W.C. (1984), *Scientific Explanation and the Causal Structure of the World*, Princeton University Press, Princeton.
van Fraassen, B. (1980), *The Scientific Image*, Clarendon Press, Oxford.
Wimsatt, W.K. Jr. and Beardsley, M.C. (1946), 'The Intentional Fallacy', in *Sewanee Review*, 54, pp. 468-88.

10 Beyond the Sciences

In the present work I have defended the methodological unity of science. Natural, social and human sciences each offer descriptions of their specific ontological domain with its respective objects and properties. However, even though the epistemic aims might be the same – be they truth or empirical adequacy – the species of explanation we seek in each field will not be. Explanations are not methods, but take their origins from the way in which we characterize the domains in question. Methods shape scientific practice; explanations take care of the content. Thus it is that we find methods to be uniform while the theories that emerge are of different kinds. And precisely because the methods have currency across the spectrum of the sciences, we face the question of how far they in fact take us in the direction of truth. Thus arises the conflict between realists and antirealists.

This conflict is philosophical in nature: the sciences cannot themselves instruct us as to whether we should be realists or antirealists. The fact that such instruction is beyond them raises the question of the boundary of scientific knowledge. Where do the demarcation lines run between ontology, science and metaphysics? Ontology and metaphysics are often conflated. Here we shall keep them apart.

Ontology is the doctrine of what exists – of all that there is in the world, whether it concern physical or mental items, meaning, values, concepts, and the principles governing them. What exists is variously studied. The sciences approach it in one way, metaphysics in another.

The most striking feature of evidence-based science – including the human and social sciences – is the entrenched conviction that our senses are the ultimate court of appeal in the determination of the truth or falsity of our assumptions. Empirical evidence is drawn from observations, experiments and readings, and is used to confirm or disconfirm scientific claims about nature, society and the human subject. This has become a constraint which, when observed, explains the authority of science and the success it has progressively enjoyed since the Greeks.

Metaphysics, for its part, articulates claims about the world that go beyond what may be ascertained through experience. Aristotle regarded metaphysics as the discipline prior to every other, concerning itself with the necessary and essential properties of what exists. Since Kant, 'metaphysics' has denoted the philosophical discipline that treats of those questions that transcend what can be established empirically. It constitutes an enquiry into fundamental reality, inaccessible to sense-experience but existing as a posited possibility, irrefutable by appeal to experience. It is well known that the positivists deemed metaphysics meaningless on the grounds that we can only have knowledge of what experience confirms. But even though metaphysics goes beyond experience, this does not lend support to the claim that its pronouncements are meaningless. Metaphysics is anything but meaningless – indeed it is what mediates meaning in scientific narratives.

What the sciences achieve

The sciences may be regarded as a highly sophisticated activity in which we engage in empirical study of things-in-themselves. Contrary to Kant, I propose that the world with which scientists engage is not restricted to that to which the senses afford immediate access but also, in equal measure, to the world comprising things-in-themselves. The sciences furnish us with an ontology predicated on readings, observations and experiments.

The natural sciences tell us that the world consists of elementary particles, atoms, fields and forces as well as planets, stars and black holes; the social sciences point to work, rules, groups, societies, states and nations as objects of research, while the human sciences focus on a world comprising the realm of the human subject with his or her mental life, culture, history, values, intentions and meanings. We are all familiar with Kant's celebrated distinction between the phenomenal world of things-as-they-appear-to-us and the noumenal world of things-in-themselves. But in fact there exists no such epistemically interesting distinction because observable things qualify as things-in-themselves too, and because science is in fact able to probe more deeply the phenomena we see with the naked eye.

It might, of course, be argued that science merely describes things as they appear to us – for elementary particles and black holes become apparent to us precisely in virtue of their observable effects. But this argument confuses the description of observable effects with that of things-in-themselves: witness the geometrical pattern in a cloud chamber imaged on a computer screen as distinct from the mass and charge of the particles which have left those tracks, despite the fact that the determination of these properties is based on the geometrical description. The difference is that between describing the effect and describing the cause – descriptions which are logically independent of each other.

Some present-day philosophers follow Kant. They deny that we can have knowledge of such things. We have seen that Bas van Fraassen maintains that only claims about observable phenomena can express well-grounded and true beliefs: it is the realm of the observable to which we have access; we have no access to that of the unobservable. And, as we have seen, this claim renders billiard balls and distant stars observable, but not atoms, fields or forces. If van Fraassen's view were correct, science would be unable to reveal the natures of things-in-themselves.

But van Fraassen's stance vis-à-vis observation is mistaken. Our belief in the existence of invisible entities derives not from their being postulated by the hypotheses reached by means of the inference to the best explanation. For we can grant van Fraassen, that were we to rely exclusively on the inference to the best explanation the conclusion would be empirically adequate long before it was true, and truth would be irrelevant to our acceptance of an explanation. Nor do we believe in unobservable things-in-themselves because they are the most probable causes of observable phenomena, as Cartwright suggests. We believe in them because we can 'observe' them, and being able to observe them is usually a sound justification for believing that they exist.

How can the unobservable be observed? The claim that it can sounds like a contradiction in terms. But we are in fact capable of observing things that we cannot see. As we have argued, there is no epistemic difference between ordinary perception and instrumental observation, for in both instances our observations rest on extensive background knowledge. There is general agreement that to recognize an identifiable object requires prior knowledge of what we are looking at, and some philosophers go so far as to maintain that all perception is theory-impregnated. This claim sometimes leads them to assert that there is nothing in our perceptual experience that remains invariant under a shift from one theory to another. But even though we need knowledge to observe that something is an item of a specific kind, such knowledge may still be objective and independent of any particular theory.

Perception is the result of sensory inputs and the cognitive processing of such inputs resulting in beliefs concerning what is observed. We have likewise established that when we have sensory awareness of something with which we are already familiar, we do not simply experience perceptual input but also acquire certain beliefs to the effect that what we are looking at is a particular kind of entity. If I see a lion, for instance, I see it in virtue of believing that what I am looking at is indeed a lion, and I believe it to be such because my sensory inputs cause just this belief in me. So it is, too, when I come across the word 'lion' in a book. I understand the word because the deliverances of my senses prompt the belief that that is the word I see, and I know what the word stands for.

What is interesting about perception is not so much its qualia aspect as its involving the acquisition of beliefs about what is perceived. But such beliefs have to be beliefs of a particular kind, not just any belief will serve us in this regard. There is no reason to doubt that many animals acquire certain beliefs when recognizing, identifying or distinguishing other animals and objects in their immediate environment. But most animals do not see things as belonging to particular categories: mice are not *mice* for the falcon. For that to be the case the falcon would have to have a concept of mice that it does not possess. So the belief we need in order to see something as an exemplar of a particular kind is a semantic belief: an individual can only perceive something to be a mouse if she holds the belief that the denomination 'mouse' denotes what she sees.

The ability to apply names appropriately, or the capacity to apply the correct predicates to things, comes to us through perceptual experience making us aware that the conditions for the application of a particular name to an item, and particular predicates to properties, are met. The acquired belief is nothing other than information to the effect that the conditions governing the application of particular words to what we experience are satisfied by what is currently in our visual field. If we so regard perception as a species of intentional behaviour, it becomes possible to generalize such a semantic account of perception to include instrumental observations. We can thus speak of observation whenever perceptual experience, with or without the aid of instrumental operations, enables an individual to grasp that the conditions for the use of a name or a predicate are fulfilled. Naturally, this presupposes a great deal of background knowledge, and to that extent our observation is indeed impregnated with theory.

So far, so good. But how does this lead to the result that the sciences address things-in-themselves? We need to be able to show that the scientist is correct in her belief that she observes the relevant invisible entities, just as we must also be able to show that we are all correct when we believe ourselves to have 'observed' visible entities.

A competent speaker is one capable of recognizing the circumstances under which names and predicates are correctly applied to what figures in his or her perceptual field. A language user would never be able to acquire the status of competent speaker if he or she failed to understand in what situations a name referred to its referent, and was unable to recognize such a situation when it obtained. When we find that these situations satisfy the criteria for the reference of the name, we normally apply the name correctly to the item in question, and this is due the fact that in the generality of cases such identification amounts to acquiring a well-founded belief that the name in fact refers – not to the situation that meets the criteria – but, more accurately, to that for which they are the criteria. Only rarely will a person err in his or her identification of such situations if he or she is to qualify as a competent language user. And in a linguistic community all are more or less well-practised users of the language.

It is no different in the case of the scientist. He or she is simply a person who is a competent user of an explicitly defined language and who uses it to communicate with others about the domain of objects for which the language, or idiom, has been evolved. There obtains a division of labour among such idioms or languages: one is used when we speak of physical entities, another when it is a matter of mental items. A researcher is characterized by his or her familiarity with the use of terms and predicates within a particular idiom or language: the physicist is familiar with the language of physics, the chemist with that of chemistry, and the economist and literary scholar with those of their respective fields. Further, she knows how they integrate with perception and instrumental procedures. In the experimental sciences it is experimental practice that fixes the reference of theoretical terms by establishing the criteria for their use. In the interpretive sciences it is other practices such as construal and interpretation that serve the same purpose. When a particular practice is introduced, when the appropriate conditions are met, the researcher possesses a well-grounded belief that she is currently observing what counts as the referent of that particular term.

We may recall that neither the descriptive theory of names nor the causal theory of reference was able to explain the use of names for natural kinds. The question was one of how the criteria for using the name ties in with the bearer of the name: how what identifies the referent connects up with the referent itself. The descriptive theory claims that natural kind terms draw their meaning from a set of predicates which state the necessary and sufficient conditions that identify the bearer of each individual name.

The theory assumes that linguistic competence involves the ability to identify the referent. But it makes no claim that the criteria have a causal connection to what the name stands for. Conversely, the causal theory of reference denies that natural kinds have a meaning specified by particular criteria determining reference. It opposes the idea that linguistic competence consists in an ability to identify what is referred to by virtue of a knowledge of some allegedly essential criteria. Reference is determined by a baptism and the causal history triggered by this event. The criterial theory places itself somewhere in between: the causal link between name and referent is determined by evidential criteria, and reference to such criteria enters into the definition of the name. The causal connection between an item and its name passes via these criteria and the item's sortal properties. We also recall that it is a logical fact that a natural entity is causally bound up with such criteria, and that, consequently, it is an ingredient in the meaning of the name that the sortal properties attaching to the bearer of the name constitute good evidence for the correct use of the name. Linguistic competence offers criteria for the identification of the referent but these may be erroneous.

It should now be clear how it is that the sciences do not simply study things-as-they-appear-to-us but things-in-themselves. First, there are those sciences that address *nominal* entities. They themselves define what it is to be a particular kind of thing, which is equivalent to saying that they determine what things are in themselves. When such definitions make reference to the properties that are accessible to sense perception it follows that we have knowledge of things-in-themselves. Second, there are the sciences that address *natural* entities. Here we have indeed established that in many cases we can only attain observable *evidence* for things, and that evidence is not identical with what it is evidence for. But we have also established that the observable evidence figures as a *logical precondition* of any reference to unobservable things. Such evidence is in fact a determinant of the referents of unobservable items because it enters into the definition of the very names of these things.

The sciences, then, give us empirical knowledge about things-in-themselves insofar as they furnish us with good evidential criteria that figure as an analytical constituent of the name. But were science to take an interest in entities and properties for whose existence no perceptual or instrumentalist evidence is available, such entities could not be the objects of empirical study. Science cannot ground claims about the world that relate to what we can neither observe with the naked eye nor discern with the aid of instruments. We would lack epistemic access to what the sortal predicates stand for. By contrast, we can be sure that atoms, molecules, vira and bacteria exist. We can be as sure of that as we are that stars, birds and other people exist. There is no epistemic distinction between the perception of macro-entities and the observation of micro-entities that validates a belief in the one but not the other.

The difference between science and metaphysics

Metaphysics begins where science ends. The distinction between science and metaphysics is a separation made not by language but by conditions of epistemic access: between what we can and cannot know. The sciences come to an end at the point at which hypotheses are empirically or intensionally underdetermined. So long as science can put well-defined questions to nature, society and human subjects and their outputs and, via sensory experience and scientific methods, is able to get well-defined answers in return, issues of what constitutes truth and falsity do not arise. But whenever the researcher is unable to elicit such answers from her material, she faces an instance of non-determination. Such situations are those where sensory experience and scientific methods are unable to determine the truth-value of alternative hypotheses. The individual hypothesis may be empirically or intensionally adequate but not necessarily true.

There are several forms of empirical (or intensional) underdetermination. The crucial point is what it is that is underdetermined. One kind of underdetermination turns on ontology and we may call it:

Global underdetermination: a hypothesis is globally underdetermined by the empirical data if there are two alternative worlds which ascribe differing truth-values to the hypothesis in question, but where the empirical data remain the same irrespective of which world is the actual world.

There are other forms of underdetermination in which it is the semantics that is empirically or intensionally underdetermined. One of these we might call:

Extensional underdetermination: a hypothesis containing theoretical terms is rendered semantically underdetermined by the language in which the evidence is expressed if the vocabulary of this language is inadequate to fix the extension of the theoretical terms.

Finally, we have a third type of underdetermination. It turns on epistemology, or rather, on the circumstance that general hypotheses of what is experienceable by us are empirically underdetermined. This type we shall call:

Local underdetermination: a hypothesis is locally underdetermined by the empirical data in a possible world if every finite set of data is inadequate to determine whether the hypothesis is true or false.

Every universal hypothesis is locally underdetermined by its data; it states more than there is a warrant for in experience, in the same way as induction *qua* method logically underdetermines its conclusion. We cannot on the basis of experience show that induction is a valid form of inference that yields true hypotheses. All inductively derived, universal claims are locally underdetermined. We looked at this form of underdetermination in chapter 6 and will not take it up again here.

The sciences yield hypotheses that may be extensionally underdetermined; metaphysics offers hypotheses that are globally underdetermined. Metaphysics pursues hypotheses which cannot be established as representative by our powers of perception and our methods. These are hypotheses about the interpretation of theories or facts where the question of the accuracy of the interpretation can never be established by reference to experimental hypotheses, sources or readings.

It happens that the sciences sometimes use divergent hypotheses to explain the same facts, hypotheses that are empirically equivalent. In such cases the truth-value of the individual hypothesis cannot be determined by empirical means since its extension is underdetermined relative to the researcher's data. We must distinguish, then, between two species of scientific hypothesis: between those that are empirically (or intensionally) decidable and those that are not. The latter are empirically underdetermined with respect to their respective extensions or, as we might say, are extensionally underdetermined.

Atoms once belonged to our hypothetical conceptions, today they are ontological existents: the boundaries between the empirically decidable and the extensionally underdetermined have not been drawn once and for all. They are concomitant on technical and intellectual advances and on the data available to the researcher. Indubitably, there will be things-in-themselves that have so far gone undiscovered, and others still that will never come to light. To the extent to which they happen just not to be empirically accessible at present – and that irrespective of whether we speak of items in the natural world or those in the mind of another person – claims regarding them will qualify as hypotheses that lie within the purview of science. And once they have become accessible to the researcher in his practical work, such things become part of science.

But often the same theory will occasion more than one interpretation of its ontological content. It is when the researcher pursues such hypotheses in areas where they are not in principle empirically or intensionally decidable that he crosses the line separating science from metaphysics. In so doing, he turns metaphysician.

Metaphysical claims

Is metaphysical knowledge at all possible? Is it possible that, while unable empirically to determine whether a hypothesis is true or false, we may yet have knowledge that just does not happen to be supported by experience? As I view the matter, the answer must be in the negative. I say this not because I would deny ontological realism, the claim to the effect that there is a mind-independent reality, or because I wish to deny the universality of language – the claim to the effect that every cognitive act is susceptible of linguistic formulation. I say it because I do not believe that it is possible to reach to what lies beyond experience.

With respect to this question also it is possible to formulate two disjunctive positions, each of which rules out the other:

Metaphysical realism contends that we can attain knowledge of fundamental reality: every ontological claim about things-in-themselves is either true or false, and we can acquire knowledge of its truth-value.

Metaphysical antirealism denies that we have any such ability to attain knowledge of things-in-themselves: no ontological claim about such things-in-themselves can be true or false, and for that very reason we can have no knowledge of its truth-value.

In this way the metaphysical realist dissociates himself from the idea that certain claims about things-in-themselves are globally underdetermined. He would rather say that metaphysical claims are *globally determinate* because they have a truth-value, and we are therefore able to gain knowledge of them. When the metaphysical realist speaks of knowledge he does not have empirical knowledge in mind. He appeals to other sources. He may align himself with reason, intuition or revelation. Metaphysical knowledge is *a priori* knowledge. By contrast, the metaphysical antirealist refuses to recognize any such source. On her judgement we can have knowledge only of things-as-they-appear-to-us. All claims relating to things-in-themselves are globally underdetermined, *i.e.* no metaphysical claim has a truth-value.

Neither of these two positions is congruent with that advocated in this book. The world is undeniably what it is, independent of our epistemic aims and interests. The world is also what it is in those aspects of its nature that lie beyond what can be established through human experience. But we can never know whether there lies something beyond experience to which we can have no access. As I conceive of the matter, should there be an inaccessible reality, one inaccessible to the methods of science, it is as it is independent of our thought, but we can never know what it is.

The position defended here lies between metaphysical realism and metaphysical antirealism. We might call it *metaphysical agnosticism* and it amounts to saying that at least some statements about things-in-themselves are true or false, and that we can have knowledge of such things (even if only of those things visible to the naked eye) while there are other analogous claims on which agnostics will disagree.

These fall into two groups, mirroring the distinction between the constructive empiricist and the entity realist:

Weak metaphysical agnosticism holds that we can have true beliefs about things-in-themselves that are not the result of an empirical investigation, but we can never know whether we possess knowledge answering to those beliefs.

Strong metaphysical agnosticism denies that we can have true beliefs about things-in-themselves unsupported by any empirical underpinnings

For the weak agnostic, metaphysical claims may be true or false even if we have no means of deciding whether they are true or false. He asserts that metaphysical claims are such that they concern things-in-themselves, but are globally underdetermined because their truth-value cannot be determined. The strong agnostic, by contrast, claims that it is impossible to have knowledge, or even true beliefs, about things that belong to the domain of metaphysics because the content of a metaphysical claim goes beyond what experience can determine. In her view, metaphysical claims are those which relate to things-in-themselves but which are *globally indeterminate* because they lack truth-value.

We may develop her view a bit further. Empirical theories are underdetermined but metaphysical ones are not at all determined by any empirical or pragmatic constraints. Some claims about things-in-themselves are concerned with things that are empirically inaccessible and these claims are therefore non-determinable. What lies beyond experience has no causal connection to experience. In the days of Democritus 'atoms' were items of metaphysical speculation; today they are not. What has changed? Today we have techniques that are able to make causal connections between perceptual phenomena and atomic events. This suggestion does not exclude that things beyond experience are able to supply beliefs about them with a determinate truth-value even though we have no knowledge of their existence. The strong agnostic maintains, however, that claims about 'atoms' were at that time neither true nor false, whereas today those claims are either true or false.

If the strong agnostic is correct, and in my view her position is the most defensible, she will be able to marshal an argument showing that claims about things-in-themselves cannot be true or false – not even by chance – unless based on empirical evidence. The strong agnostic is an ontological realist: she maintains that facts about the world are not contingent on what it is possible for human cognizers to ascertain at any point in time. But that is naturally not enough. She needs two further premises.

One is to the effect that every linguistic representation of the world builds on a convention. Linguistic signs refer to nothing at all in advance of a determination of their referents. Words and sentences must be invariantly linked with the objects and facts that they represent through an established causal connection.

The other concerns the circumstance that only *signs*, *i.e.* individual statements, claims and beliefs are possible truth-bearers and that facts are possible truth-makers. It is individual statements, claims and beliefs that are true or false. Truth is a property of signs, not of the world. Nor is truth a property of sentence types but only of token utterances and thoughts. Only in this way are we able to explain the fact that the same sentence type can vary in truth-value. The connection between an actual truth-maker and a relevant truth-bearer is called a *truth relation* and truth-makers and truth-bearers are *relata* in this semantic relation. Consequently, an individual sentence such as 'There are lions in Africa' is true if and only if it is a fact that there are lions in Africa.

With these two premises in place, the strong agnostic is in a position to present her argument: A sign is true because it stands in a particular relation to the relevant fact that makes it true. The truth relation obtains between two relata, for example between a sentence and a fact and this relation between truth-bearer and truth-maker is normally taken to be internal. Philosophers simply tacitly assume that that is the way it is. Few have taken the trouble to offer a justification. That the truth relation is internal is to be understood as meaning that it is impossible for a truth-maker to exist without the corresponding truth-bearer being true. The truth-bearer is automatically true if the truth-maker exists.

It is precisely this that the strong agnostic denies. She holds that it is possible that there are truth-makers which do not make anything true. This implies that such facts are merely potential truth-makers. The background to this lies in the celebrated dispute between J.L. Austin and Peter Strawson as to whether facts are distinct from true statements. But there can be no doubt that they are. To maintain otherwise is to make reality linguistic and to confuse linguistic representation with what language represents. We *see* facts: the computer is 'eye-catchingly' there on the desk whenever I use it. And had I no concept of computer, desk, things etc. I should still have seen it. Facts cannot act as truth-makers unless we can identify them independently of language as non-linguistic facts. A correlation between language and reality can exist only if the correlata are independently identifiable. If it were so that sentence types constrained fact types, the constitution of the world would be determined by language. In the beginning there was language, then came the world. But everything we know about the evolution of the universe and human beings tells us that the reverse is the case.

The argument remains as yet unfinished. To complete it we need to adduce the following considerations. An individual fact that furnishes an utterance with a truth-value has objective existence, independent of the sign. Such a fact, however, cannot make the utterance true if it cannot be individuated independently of the relevant truth-bearer. It must be identifiable as a fact generically describable by the sentence type exemplified by the utterance. This is only possible in those cases where we first have a *descriptive correlation* between a particular sentence type and a particular fact type – a correlation that determines to which facts such sentences conventionally refer. In other words, we can say that an individual sentence can stand in relation to an individual fact only when we have had the opportunity to identify both sentence type and fact type in isolation from each other. A descriptive correlation – like any other – can obtain only if the items correlated are known to exist independently of each other. The relevant fact type must thus be available to our cognitive powers for such a correlation to exist, and a matching causal nexus between referent and the referring expression established. Only with this connection in place can a truth relation between a particular linguistic expression and a particular ontological referent obtain.

If this argument is watertight, it has wide-reaching implications for metaphysics. It makes it impossible that we should have metaphysical knowledge. It is possible for a fact to exist without making a sentence true, if its being a fact of a particular type has gone unrecognized. It means that a token sentence such as 'God is omnipresent' is neither true nor false; nor would it be were God even to exist, unless it had been established that the term 'God' refers to a being that is or is not omnipresent. We must have other means than linguistic means by which to identify God if we are to say that we have knowledge of God. That we have the word 'God' is not enough to establish the existence of a referent – meaning is not sufficient to secure reference.

In the history of philosophy *a priori* arguments have often been attempted and rationalists in particular have used them to try to justify synthetic claims. But before a synthetic sentence is able to state something true about the world, we need to be able to establish a descriptive correlation between a sentence type and a particular fact type. This can only be achieved by empirical means. *A priori* arguments cannot give us knowledge about things-in-themselves not revealed by experience.

Nor is there help to be sought in intuition or personal revelations. Neither of these psychological states has established itself as a reliable epistemic resource. With respect to perception we know that our senses sometimes err because commonsense and science have developed both simple and sophisticated methods to detect error and correct it. But as long as we have no methods to tell us how intuitions and revelations are to be refuted – and with what means – such postulated powers cannot be said to have reliable credentials and so cannot furnish us with knowledge of metaphysical fact.

The experience-based identification of non-linguistic facts explains the success of science as purveyor of knowledge. Science enjoyed success from the moment scientists understood that truth-making facts are not identified through the use of language, but through the use of the senses. Metaphysics, by contrast, is not concerned with claims that are globally underdetermined but with those that are globally indeterminate. Such claims are neither true nor false. Their acceptability is determined by their internal consistency, simplicity and their coherence with other interpretations. This explains why metaphysics has never succeeded as a purveyor of knowledge. There is nothing there that can be known.

However, we ought not, on that account, be silent about what lies beyond the bounds of empirical enquiry. Philosophical argument has figured prominently in the present work and metaphysics is precisely a philosophical account of the world, human beings and meaning. Metaphysics is as little meaningless as are norms, values and fictions. But it is a mistake to judge it according to the premises of empiricism. We engage in metaphysics whenever we justify a choice between divergent globally underdetermined hypotheses, whenever we attempt to give our knowledge form and perspective, and whenever we endeavour to align our epistemic values with political, religious, moral and aesthetic values. We can conceive of space and time as absolute or relative, or the atomic world as a many-world or Bohmian, and we may hold that the Big Bang was brought into being by a creator or by random quantum fluctuations in a vacuum state. By the same token, we are capable of conceiving of the mind as mental or material, of meaning and values as objective or subjective, and of subscribing to determinism or to free will. No matter what our respective beliefs, this manifold of scenarios offers a broad canvas against which, *qua* human agents, we seek to plumb the knowledge we have of things-in-themselves and integrate it into a unified whole. It is immaterial whether the subject matters concern the things-in-themselves that natural science discloses to us, or the questions that preoccupy the human sciences. It is in metaphysics that we seek an ultimate and coherent and comprehensive interpretation of life – the more coherent and comprehensive, the more persuasive.

Metaphysics, then, is not knowledge but the aspiration to probe how our knowledge relates to our life.

Index

abduction, 84, 86, 91, 93, 94, 95, 110
abstraction, 8, 58, 115, 141, 143, 144
Andersen, H.C., 81, 82
Anscombe, Elisabeth, 36, 45, 54
anti-intentionalism, 189, 190, 194
antirealism
    epistemological, 170
    ontological, 170
    scientific, 173
    semantic, 170
Aristotle, 4, 30, 149, 177, 178, 201
Armstrong, David M., 122, 123, 141
Austin, John L., 212

Beardsley, Monroe, 189, 200
Bell, J.D., 177
Bird, Alexander, 172, 176, 199
Bloor, David, 185, 186, 187, 199
Bohr, Niels, 60, 62, 63, 126, 157, 164
Brahe, Tycho, 101
bridge law, 12, 13, 14, 17, 18

Cartwright, Nancy, 126, 127, 130, 141, 179, 181, 182, 184, 199, 203
causation, 128, 129, 133, 141
cause, 31, 32, 33, 34, 37, 39, 40, 41, 42, 44, 56, 57, 80, 88, 90, 91, 93, 94, 111, 116, 128, 129, 181, 182, 183, 184, 198, 202, 204
common cause principle, 182, 183
Copernicus, 101, 102
criteria
    sortal, 199

critical rationalism, 100, 102, 104
Currie, Gregory, 190, 194, 199

Davidson, Donald, 37, 45, 82
Dilthey, Wilhelm, 47
Dretske, F., 122

Einstein, Albert, 59, 162, 167, 172
Elster, Jon, 37, 48
emergentism, 11, 17, 19
empirical adequacy, 9, 44, 94, 109, 147, 175, 176, 177, 178, 179, 201
empirical equivalence, 177
empiricism
    classical, 20, 51, 174, 175, 214
    constructive, 174, 175
epistemic goal, 22, 43, 44, 84, 92, 95, 105, 107, 109, 111, 113, 114, 147, 198
essence
    nominal, 68, 69, 70, 72, 73, 75, 78, 80
    real, 68, 69, 70, 71, 73, 75, 76, 77, 78
explanation
    answer to a how-question, 43
    answer to a what-question, 43
    answer to a when-question, 43
    answer to a why-question, 28, 31, 32, 33, 36, 40, 41, 43, 44, 45
    causal, 24, 25, 31, 32, 33, 36, 37, 38, 41, 42, 45, 48, 55, 105, 107, 113, 153, 181
    functional, 32, 33, 34, 35
    functionalist, 32, 33, 34, 35
    inference to the best, 83, 84, 86,

94, 95, 97, 107, 112, 166, 172, 175, 176, 177, 178, 181, 184, 203
  intentional, 24, 32, 36, 39, 40, 42, 43, 44, 45, 49
  interpretive, 24, 32, 40, 41, 42, 44, 53, 55, 81, 106, 107, 109, 110, 139
  material, 43
  nomic, 31, 32, 33, 34, 130
  structural, 43

fact
  natural, 82
  nominal, 82
  the criterial position, 72, 75, 158, 206
  the descriptive position, 65, 66, 67, 69, 72, 75, 205
  the essentialist position, 68, 69, 72
falsificationism
  simple, 101
  sophisticated, 101
Feyerabend, Paul, vi, 62
Fish, Stanley, 192, 199
Fodor, Jerry, 17, 18, 19, 22
Freud, Sigmund, 49, 146, 162, 165
Føllesdal, Dagfinn, 107, 105, 113

Gadamer, Hans-Georg, 5, 6, 10, 109
Giere, Ronald, 146, 167
Gould, Stephen J., 34, 45
Greimas, J. A., 155
Grünbaum, Adolf, 41, 46

Habermas, Jürgen, 6, 10, 105
Hacking, Ian, 179, 180, 181, 199
Hamilton, William R., 131
Harman, Gilbert, 176
Heidegger, Martin, 5, 190
Heisenberg, Werner, 158, 164
Hempel, Carl G., 25, 26, 27, 46, 116

hermeneutics, 4, 5, 50, 105, 106, 107, 108, 109, 113, 138
Hirsch, E.D., 190, 199
holism, 11, 17, 50, 133, 134
  methodological, 133
  ontological, 133
Homer, 163
Hume, David, 88, 89, 117
hypothetical intentionalism, 189, 190, 191, 192, 198

Ibsen, Henrik, 108
idealization, 8, 115, 141, 142, 144
idiographic interpretation, 137
indeterminacy
  global, 211, 214
induction, 9, 84, 85, 86, 87, 88, 89, 90, 91, 92, 93, 100, 104, 117, 171, 208
instrumentalism, 173, 174, 188, 198
intensional adequacy, 198
intentionalism, 189
interpretation, 2, 4, 5, 6, 9, 24, 32, 41, 42, 44, 47, 48, 49, 50, 51, 52, 53, 54, 55, 56, 57, 59, 60, 61, 62, 63, 64, 82, 93, 103, 110, 108, 109, 111, 112, 113, 114, 115, 116, 132, 134, 137, 138, 141, 122, 124, 141, 143, 145, 146, 147, 148, 152, 154, 162, 186, 187, 189, 192, 193, 194, 195, 196, 197, 198, 199, 200, 205, 208, 209, 214
interpretive instrumentalism, 189, 190

Kant, Immanuel, vi, 77, 201, 202, 203
Kemeny, John, 14, 15, 22
Kim, Jaegwon, 18, 19, 23
Kneale, W.C, 121, 122, 141
Kripke, Saul, 69, 70, 73, 75, 82
Kuhn, Thomas, 130, 160, 161, 162,

163, 164, 165, 167

Lagrange, Joseph L., 131
Lakatos, Imre, 164, 165, 166, 167
language
    observational, 20
    theoretical, 20
Lassen, K., 150
Latour, Bruno, 187, 199
law
    as linguistic rule, 44, 57, 111, 130, 131, 134, 135, 138, 142, 148, 149, 153, 154, 155, 167, 168,
    causal, 18, 26, 32, 121, 128, 129, 131, 132, 137, 138, 139, 159
    ceteris paribus, 126, 127, 128, 131, 134, 138, 139, 152, 153, 154, 159
    constructivist position, 123
    deconstructivist position, 125
    fundamental, 125, 126, 127, 128, 130, 131, 135, 153, 181
    maximalist position, 121, 123
    minimalist position, 116
    phenomenological, 126, 127, 128, 135, 153
    social, 12, 132, 133, 134, 135
Levinson, Jerrold, 190, 191, 192, 197, 199, 200
Lewis, David, 123, 124, 141
Lister, Joseph, 91
Locke, John, 68

Marx, Karl, 45, 133, 134, 135, 146, 165
meaning
    experiential, 50, 51, 52
    linguistic, 50, 51, 146, 147, 161, 162, 170
    literal, 40, 41, 61, 110, 190, 191, 192
    literary, 40, 191, 192, 193, 195, 196, 198, 199
    metaphorical, 40, 41, 61, 92
    textual, 190, 192, 193, 196, 197, 198, 199
Melden, Abraham L., 37, 46
Merton, Robert, 35
metaphysics
    agnosticism, 210
    antirealism, 210
    realism, 209, 210
method
    general, 83, 84, 85, 105
    hermeneutical, 47, 83, 104, 105, 106, 109
    hypothetico-deductive, 47, 83, 85, 99, 100, 102, 103, 104, 105, 107
    specific, 83, 84
methodological prescription, 22, 98, 111, 112
methodology, 5, 44, 62, 106, 112, 164, 167
misinterpretation, 109, 112, 113
model, 25, 26, 27, 59, 116, 126, 127, 130, 131, 134, 140, 142, 146, 147, 148, 149, 150, 151, 152, 153, 154, 155, 156, 157, 158, 159, 161, 164, 169, 174, 177, 197

Nagel, Ernst, 12, 13, 23
natural kind, 4, 16, 17, 18, 51, 65, 66, 67, 68, 69, 70, 71, 72, 74, 75, 76, 79, 80, 86, 115, 128, 139, 142, 164, 205
Newton, Isaac, 7, 13, 57, 58, 114, 124, 125, 126, 127, 130, 131, 132, 134, 142, 144, 148, 152, 161, 162, 165, 167, 172

observation, 3, 9, 20, 40, 38, 45, 60, 64, 65, 73, 87, 100, 103, 110, 117, 151, 178, 203, 204, 207
ontological principle, 92, 93, 94, 98, 112, 161

ontology, 47, 107, 123, 201, 202, 207
Oppenheim, Paul, 14, 15, 16, 22, 23, 116

Panofsky, Erwin, 139, 141
paradigm, vi, 160, 161, 162, 163, 164, 165, 166
Pasteur, Louis, 91
perception, 20, 49, 52, 59, 65, 203, 204, 205, 206, 208, 213
Perrin, Jean B., 182
phenomenalism, 173
Popper, Karl, R., 99, 100, 102, 104, 113, 121, 122, 141
positivism, vi, vii, 4, 10, 20, 21
properties
  accidental, 78
  sortal, 74, 75, 76, 77, 78, 80, 81, 87, 143, 144, 158, 183, 206
psychoanalysis, 41, 46, 138, 142, 146, 154, 155, 189, 197
Putnam, Hilary, 16, 23, 69, 70, 71, 73, 75, 82, 171, 172, 179

Ramsey, Frank P., 119, 120, 123, 124, 141
rationalism, 51, 100
readings, 9, 61, 111, 112, 141, 156, 201, 202, 208
realism
  entity, 179, 180, 181, 189, 196, 199
  epistemological, 169
  ontological, 169, 209
  scientific, 123, 168, 170, 171, 172, 178, 188
  semantic, 169, 174
  theory, 179, 180
reduction
  concept, 12
  law, 12
reductionism

eliminative, 16
non-eliminative, 16
ontological, 16, 17, 19
the cosmological argument, 15
the epistemological argument, 20
the logical argument, 11
Reichenbach, Hans, 99, 182, 183
relativism, 6, 185, 186, 198
research programme, 164, 165, 166
Ricoeur, Paul, 41, 46
rule
  constitutive, 136
  correspondence, 145
  regulative, 136
  social, 135, 136, 137
Rutherford, Ernst, 156, 164
Ryle, Gilbert, 36, 46

Schrödinger, Erwin, 158, 165, 181
semantic definitions, 92
Semmelweis, Ignaz P., 90, 91, 92
sentence
  analytic, 37, 77, 78, 79, 80
  synthetic, 77, 78, 79, 213
social constructivism, vi, 184, 185, 186, 187, 188
Stinchcombe, Arthur, 36
Strawson, Peter E., 89, 212
structuralism, 7, 142, 155, 189, 197
supervenience, 18, 19, 23
Suppe, Frederick, 146, 167
Suppes, Patrick, 156, 167
Svensmark, Henrik, 150

Taylor, Charles, 49, 50, 51, 52, 63
Tolhurst, William, 190, 191, 192
Tooley, Michael, 122
truth, vi, 3, 5, 9, 20, 21, 22, 30, 38, 44, 45, 77, 82, 89, 95, 114, 126, 127, 147, 167, 169, 171, 172, 174, 175, 176, 177, 178, 181, 184, 185, 187, 193, 196, 198,

201, 203, 207, 211, 212, 213
truth relation, 211, 212, 213

underdetermination
  behavioural, 110
  empirical, 14, 61, 95, 125, 166, 177, 208
  extensional, 207
  global, 207, 208, 209, 210, 214
  intensional, 110, 118, 121, 208, 212, 214
  local, 208
understanding, 1, 3, 5, 6, 7, 8, 10, 21, 22, 24, 25, 27, 29, 38, 46, 47, 48, 49, 52, 53, 54, 55, 56, 60, 61, 63, 64, 67, 72, 91, 103, 105, 106, 108, 109, 110, 112, 113, 114, 127, 136, 137, 138, 140, 141, 143, 144, 147, 152, 153, 155, 158, 159, 160, 165, 187, 188, 190, 191, 193, 194, 196, 198
  horizon of, 6

van Fraassen, Bas, 27, 28, 31, 43, 46, 59, 63, 146, 147, 167, 171, 174, 175, 176, 177, 178, 179, 183, 184, 194, 200, 203
von Wright, Georg H., 38, 46
Wimsatt, W.K., 189, 200
Wittgenstein, Ludwig, 110
Woolgar, Steve, 187, 199